Ocean Thermohaline Circulation

Climate Feedbacks and Global Change

Team Trex

Published by NobleTrex Press

For permissions and other inquiries, write to:
P.O. Box 3132, Framingham, MA 01701, USA

Contents

Introduction

If you were to stand on the western coast of Scotland in mid-January, watching the gray Atlantic rollers smash against the Hebrides, you might find it hard to believe that you are, geographically speaking, sharing a latitude with Hudson Bay. You are roughly as far north as the polar bears of Churchill, Manitoba. Yet, while the Canadians are digging out from meters of snow in temperatures that can freeze exposed skin in minutes, the Scots are likely wearing raincoats, grumbling about the drizzle, and perhaps tending to winter-hardy palm trees in local gardens.

This atmospheric inequality is one of the great fortunate accidents of human history. Europe-a continent that has hosted empires, revolutions, and the birth of the Industrial Age-exists in a bubble of thermal privilege. It is artificially warmed, kept on life support by a massive, invisible planetary heating system that pumps staggering amounts of energy northward, day in and day out, without fail.

We are talking about heat on a scale that defies human comprehension. To match the thermal energy delivered to the North Atlantic by the ocean every year, you would need to run a million large power plants simultaneously. Or, to put it another way, the ocean moves more heat toward the poles than is consumed by all of humanity combined, multiplied a hundred times over.

This heat does not move by magic. It travels on water. It is the result of the Ocean Thermohaline Circulation-the "Great Conveyor Belt," as the late, brilliant geochemist Wally Broecker famously dubbed it. It is a system of currents that connects the surface of the sea to the crushing darkness of the abyss, binding the Pacific, Indian, and Atlantic oceans into a single, writhing, global loop. It is the heartbeat of the planet.

And it is slowing down.

For decades, the idea that ocean currents could change on a human timescale was dismissed as science fiction. The ocean was viewed as the "flywheel" of the climate system-immense, sluggish, and immutable. We thought that while the atmosphere might be flighty and chaotic, the deep ocean was the steady hand on the tiller. We were wrong. As we have peered closer at the geological archives buried in mud and ice, and as we have deployed armies of robotic sensors into the deep, a different picture has emerged. The ocean is dynamic. It is sensitive. And it has a tipping point.

This book is an exploration of that planetary heartbeat. It is a story about how the weight of water shapes the fate of nations, and how a delicate balance between salt and heat has kept our climate stable for the last ten thousand years-the very window in which human civilization arose. It is also a detective story, piecing together clues from the ocean's past to understand a future that is becoming increasingly uncertain.

***

To understand the stakes, we must first appreciate the machinery. The engine of this great circulation lies not in the sun-drenched tropics, as intuition might suggest, but in the hostile, freezing corners of the high latitudes. In the Nordic Seas and the waters surrounding Antarctica, a violent physical process takes place that powers the world. Here, wind and cold strip heat away from the surface water, making it dense. As ice forms, it rejects salt, leaving the remaining water even saltier and heavier. Eventually, this water becomes so dense that it can no longer float. It plunges.

This vertical crash-billions of tons of water sinking miles deep into the abyss-is the piston that drives the engine. It pulls warm surface water northward to replace what sank, drawing heat from the tropics toward the poles. It is this mechanism that keeps London mild while Labrador freezes. It is this mechanism that regulates the rainfall patterns over the Amazon and the monsoons of India.

But this engine runs on a specific fuel: density. And density is a function of two things: temperature (thermo) and salt (haline). This is where the story gets complicated, and where our current moment in history becomes critical.

We are currently running a dangerous experiment with the machinery of the Earth. By warming the atmosphere, we are melting the great ice sheets of Greenland and the Arctic. We are altering the global hydrological cycle, causing more rain to fall over the high north. All of this pours fresh, buoyant water into the very regions where the ocean is supposed to sink. Fresh water is the enemy of the conveyor belt; it acts like a lid, a buoyancy cap that refuses to go down. If you pour enough fresh water onto the North Atlantic, you jam the piston. The circulation slows. If you push it hard enough, the physics dictates that it must stop entirely.

This is not merely a theoretical concern derived from equations on a chalkboard. It has happened before.

As we will see when we look back at the "Ghosts of Climates Past," the Earth carries the scars of previous shutdowns. We know from analyzing air bubbles trapped in ancient Greenland ice cores that the climate is capable of flickering like a faulty lightbulb. During the last deglaciation, a massive pulse of fresh water from a glacial lake in North America flooded the North Atlantic. The response was instantaneous and catastrophic. The heat pump failed. The Northern Hemisphere plunged back into near-glacial cold within a matter of decades-a geological blink of an eye. This event, known as the Younger Dryas, serves as a stark reminder: the climate system is not a linear dial that we gently turn up or down. It is a complex, beastly thing with switches, thresholds, and surprises.

***

Our journey into this subject requires us to abandon the intuitive world of human scales and enter the realm of geophysical fluid dynamics, yet the concepts are surprisingly grounded in everyday physics. If you have ever watched cream swirl into coffee or felt the draft from an open window, you have witnessed the fundamental forces that drive the ocean. The difference is merely one of magnitude.

We will begin by tracing the architecture of these currents, moving from the wind-whipped surface where the Gulf Stream meanders like a river, down to the abyssal plains where water moves with the slowness of a glacier. We will meet the pioneers of oceanography, figures like Henry Stommel, who, working with nothing more than pencil and paper, deduced the existence of deep opposing currents years before anyone could measure them.

The narrative will then take us to the "Gateways to the Abyss," the specific geographic chokepoints where the ocean takes its plunge. These are dramatic, dangerous places. In the Denmark Strait, between Iceland and Greenland, there exists a waterfall entirely underwater that dwarfs the Angel Falls or Victoria Falls. It creates a cascade of cold water plummeting thousands of meters into the deep, a silent torrent that helps drive the global circulation.

But the system is fighting a constant battle for buoyancy. We will examine the delicate interplay between the "Salt-Advection Loop"-a positive feedback mechanism that sustains the current-and the influx of fresh water that threatens to disrupt it. This brings us to the concept of hysteresis, a term borrowed from physics that explains why some systems are easy to break but nearly impossible to fix. If the AMOC (Atlantic Meridional Overturning Circulation) were to collapse, simply removing the fresh water wouldn't restart it. The system has a memory. It would stay broken, perhaps for centuries or millennia, locked in a new, stagnant equilibrium.

This brings us to the question that keeps oceanographers awake at night: How close are we to the edge?

For years, this question was unanswerable. The ocean is opaque to satellites, which can only see the skin of the surface. To measure the deep heartbeat, you need to go there. We will look at the heroic engineering efforts of the last two decades, such as the RAPID array-a string of sensors moored across the entire width of the Atlantic Ocean. These instruments have provided us with our first continuous "ECG" of the ocean's circulation. The data they have returned is both fascinating and sobering. The variability is higher than we thought, and the signs of a slowdown are beginning to emerge from the noise.

***

Why should the average person, living perhaps hundreds of miles from the nearest coast, care about the turnover rate of the North Atlantic?

Because the ocean is the flywheel of civilization. A collapse of this circulation would not look like the Hollywood dramatization of instant freezing, but the reality would be no less devastating for the infrastructure of human society. Models suggest that a shutdown would severely disrupt the rains that water the crops of West Africa and India, putting billions at risk of food insecurity. It would cause a dynamic rise in sea levels along the Eastern Seaboard of the United States, piling water up against New York and Boston as the Gulf Stream slackens. It would alter the storm tracks that steer hurricanes, and strangely, while the north cooled, it would trap heat in the Southern Hemisphere, accelerating the warming of the rest of the globe.

Furthermore, the overturning circulation is our primary garbage disposal for carbon dioxide. The deep water formation carries distinct chemical fingerprints-human-made carbon absorbed from the atmosphere and locked away in the deep for centuries. If the conveyor stops, that service ends. The ocean's ability to buffer us from our own emissions would diminish, causing atmospheric warming to spike even faster.

We are standing at a unique moment in geologic history. We are the first species capable of understanding the planetary machinery that supports us, and simultaneously the first species capable of breaking it. The story of the ocean's overturning circulation is the story of a battle between the immense inertia of the natural world and the accelerating pressure of human influence.

The water that is sinking in the Labrador Sea today will not see the surface again for a thousand years. It will travel south, cross the equator, loop around Antarctica, and surface perhaps in the North Pacific in the year 3025. What kind of world will it return to? That depends entirely on the stability of the system we are about to explore.

So, let us dive in. We will leave the solid ground behind and follow the water-from the sunlit tropics to the crushing dark, from the salt-spray of the surface to the silence of the bottom. It is a journey through the veins and arteries of our living planet, revealing a system that is at once powerful, beautiful, and terrifyingly fragile.

Chapter 1

The Planetary Heartbeat

Beneath the familiar chaos of surface waves lies a slow, silent current of immense power, moving water from the Arctic to the Antarctic and back again. This global conveyor belt acts as Earth's thermostat, redistributing solar energy to keep the tropics cool and the poles habitable, yet its stability is far more fragile than its size suggests.

1.1 Broecker's Great Conveyor

Scientific revolutions do not always arrive with a new instrument or a new equation. Sometimes they arrive as a picture you can hold in your head.

By the late twentieth century, oceanographers had already mapped the oceans in astonishing detail. Ships crossed the Atlantic and Pacific along repeating transects, lowering bottles and sensors into the water column. They returned with vertical profiles of temperature and salinity, and with them a new kind of fingerprinting: the ocean is not one uniform body, but a layered archive of distinct water masses. A parcel of water formed in one region carries a characteristic combination of warmth, saltiness, dissolved gases, and chemical tracers, and that signature can be followed far from its birthplace. Piece by piece, these surveys turned the deep ocean from an imagined darkness into something like a three-dimensional weather map.

But the very richness of those data posed a problem. Temperature—salinity diagrams, contour sections, and tracer distributions were precise, yet not easily grasped as a single story. If you wanted to explain to a non-specialist why a change in the North Atlantic could matter for rainfall in the tropics or the carbon content of the deep Pacific, you faced a challenge of translation. The science was correct, but it did *travel*.

Wallace "Wally" Broecker, a geochemist with a gift for synthesis, solved that problem in the way talented explainers often do: by compressing complexity into a metaphor that was simple enough to remember, but sturdy enough to guide thought. In the 1980s, he popularized an image of the ocean's global overturning as a single looping circuit: warm surface waters moving poleward, cooling and becoming denser in the far North Atlantic, sinking to depth, and eventually returning through the Indian and Pacific before rising again and completing the loop. He called it the "Great Ocean Conveyor"—a phrase that made the ocean sound less like a featureless basin and more like a machine with moving parts.

The power of the metaphor lay partly in its audacity. The ocean is not literally a conveyor belt. It does not run on rollers; it does not move at a single speed; it does not carry identical parcels in a tidy line. Yet the metaphor captured something that had been difficult to say in ordinary language: the ocean is not merely *connected* by currents along coastlines, but *interlocked* by transformations that link distant basins through the deep sea. The surface ocean exchanges heat and freshwater with the atmosphere; those exchanges change density; density differences drive sinking and spreading; and the deep ocean, though slow, is vast enough to matter for climate and carbon. The conveyor was a way to speak about a global coupling between physics and chemistry, between circulation and climate, without asking the reader to learn a new technical vocabulary.

A second reason the image caught fire was timing. In climate science, the late 1970s and 1980s were years when a new sensibility was taking hold: climate might not always change gently. Paleoclimate records—especially those extracted from Greenland ice—were revealing abrupt shifts in temperature on timescales of decades or even years during the last glacial period. The Younger Dryas, a cold snap that interrupted the deglaciation roughly 12,900 years ago and ended around 11,700 years ago, became a particularly vivid example. It looked, in the archives, like a climate system capable of snapping between states. Scientists argued over mechanisms, but the North Atlantic kept reappearing in the story, because it is one of the few places where surface waters reliably become dense enough to sink to great depth.

Broecker's genius was to connect these threads: the detailed hydrographic picture of distinct deep waters; the idea that the Atlantic had a special role in producing them; and the paleoclimate evidence that the climate system could reorganize abruptly. Once you drew the ocean as a loop with a crucial sinking branch in the North Atlantic, you had a natural place to insert a switch. If the sinking weakened, the whole loop could slow; if the loop slowed, the distribution of heat, salt, carbon, and nutrients could change; and those changes could feed back onto climate. A sprawling set of observations was turned into a narrative with tension: an engine room, a lever, a possibility of failure.

A metaphor this successful can make its creator seem like a lone inventor, but it is better understood as a culmination. Oceanographers had long discussed "thermohaline circulation," a phrase that points to temperature (*thermo*) and salt (*haline*) as key controls on density. Earlier schematic diagrams of meridional overturning existed, and earlier theoretical work had shown how sensitive a coupled temperature—salinity system could be. Henry Stommel, in a classic 1961 model, demonstrated that a simplified two-box ocean could admit more than one stable circulation regime under the same external forcing. In ordinary language: even if the winds and the sun stayed the same, the circulation might settle into either a vigorous overturning state or a sluggish one, depending on history. The model was stripped to the bone, but it introduced an unsettling possibility: hysteresis, the idea that the path into a state is not necessarily the path out of it.

Broecker took such ideas and gave them a face. He did not ask the reader to picture equations; he asked them to picture plumbing. And plumbing is something people intuitively understand. If you block a pipe, the flow changes everywhere. If you dilute a dense brine with freshwater, it may stop sinking. If a pump slows, the whole circuit becomes lethargic.

That plumbing intuition hinges on one physical fact: seawater density depends on both temperature and salinity. Cold water tends to be denser than warm water. Salty water tends to be denser than fresh water. In the subtropics, evaporation leaves the surface relatively salty; in rainy regions and near melting ice, the surface can freshen. In high latitudes, the ocean also loses heat to the atmosphere, especially in winter storms. The North Atlantic is unusual because it combines routes that bring in warm, salty water at the surface with intense winter cooling. Salty water, preconditioned in lower latitudes, arrives and then is chilled; cold *and* salty is a recipe for sinking.

Once water sinks, it does not drop straight to the seafloor like a stone. It spreads along density surfaces, sliding beneath lighter waters and above denser ones, and it can be steered by undersea topography. Yet, averaged over time, the effect resembles a vertical exchange: surface waters transformed into deep waters, deep waters eventually returning to the surface elsewhere. This is the overturning circulation: not a single current, but a statistical balance of transformations and compensating flows.

Broecker's diagram made one further move: it treated the Atlantic as the main site where deep waters form and descend, and the Indian and Pacific as the basins where those deep waters gradually return upward and back toward the surface. That return is not magic; it is enabled by mixing and by interactions with the winds, especially in the Southern Ocean, where strong westerlies and the geometry of the Antarctic Circumpolar Current help bring deep waters back toward the surface. But the conveyor metaphor allowed readers to grasp the broad logic: sink in one place, rise in another, and link the whole planet in between.

A concrete anecdote helps explain why this mattered. In the early days of radiocarbon oceanography, Broecker and others used the distribution of carbon-14 to infer ventilation ages of deep waters. The deep Pacific, in this view, looked "old": it had been out of contact with the atmosphere for a long time, because radiocarbon decays. The deep North Atlantic, by contrast, looked younger. One could tell a coherent story: newly formed deep water enters the abyss in the North Atlantic, spreads into other basins, and by the time it reaches the deep Pacific it has aged. This is not merely a curiosity about isotopes. It implies that changes in the formation rate of North Atlantic deep waters would alter how quickly the deep ocean can absorb heat and carbon from the atmosphere. It also implies that the chemical state of the ocean interior—including oxygen levels important for marine life—is tied to a slow, global-scale circulation pattern.

The metaphor also clarified, at a glance, why the Atlantic was climatically special. It is tempting to equate the Atlantic's role with the Gulf Stream, because the Gulf Stream is fast, visible in satellite images, and famous. But Broecker's conveyor aimed at something deeper and slower: the overturning component that carries surface waters northward, transforms them to depth, and returns a compensating flow southward in the deep Atlantic. This deep limb does not make headlines the way a western boundary current does, yet it is part of a system that rearranges heat and salt over enormous distances. The result is not a gentle background hum; it can be felt in the contrast between western Europe and other regions at similar latitudes.

Still, a successful metaphor comes with a cost. Every schematic hides as much as it reveals. The conveyor picture suggests a single, continuous ribbon of flow, as if the ocean were a well-marked highway. The real ocean is more like a crowded city with side streets, detours, and eddies that send water parcels on many different routes. Water can recirculate within a basin rather than joining the global loop. Multiple deep water masses coexist: North Atlantic Deep Water is not the same as Antarctic Bottom Water, and the boundary between them shifts. Upwelling occurs in several places and by more than one mechanism. Even the notion of a single "speed" for the conveyor can mislead, because different parts of the circulation operate on different timescales: surface currents can move at meters per second, while deep interior flows may creep at centimeters per second or less.

Yet those cautions do not diminish what Broecker achieved. A schematic that oversimplifies can still be scientifically productive if it puts the right causal relationships into the foreground. The conveyor image taught generations of students and citizens to ask a specific kind of question: what happens to climate when you perturb the density structure of the North Atlantic? It also gave policymakers and journalists a handle, for better or worse, on an otherwise technical subject. When people talk about the "AMOC" today—the Atlantic Meridional Overturning Circulation, a more precise modern term—they often have Broecker's picture in mind, even if they have never seen the original drawing.

The most profound influence of the conveyor metaphor may be psychological. Climate change is often discussed in terms of *gradual* forcing: more greenhouse gases, more heat retained, more warming. The ocean conveyor introduces a second mental model: the possibility of thresholds. Not a guaranteed collapse, not a Hollywood instant ice age, but a system whose strength depends on a delicate balance of heat loss, freshwater input, wind forcing, and internal mixing. The balance can shift. The language of "tipping points" can be abused, but it reflects a legitimate scientific idea: some components of the Earth system can respond nonlinearly to steady pressure.

This is why Broecker so often returned to the Younger Dryas. That event, whatever its ultimate trigger, showed that the North Atlantic region could undergo a dramatic reorganization. A commonly discussed mechanism involves freshwater: large inputs from melting ice sheets or rerouted glacial lakes could freshen the surface, making it less dense and less prone to sink even when cooled. In the conveyor metaphor, that is like diluting the heavy fluid feeding a downflow pipe. If sinking slows, less warm water is drawn northward at the surface, potentially cooling the region further. At the same time, the redistribution of salt can change, sometimes amplifying the original perturbation. The details are complex and still actively studied, but the metaphor kept the spotlight on a central suspect: changes in the high-latitude North Atlantic that reverberate through a global loop.

Placed on a simple world map, the conveyor looks almost inevitable: a red surface ribbon flowing north in the Atlantic, plunging to blue in the subpolar North Atlantic, returning southward at depth, wrapping around Antarctica, and branching into the Indian and Pacific before rising again. That image has been printed in textbooks, museum exhibits, and government reports. It has been mocked and revised and debated. But it has also done what the best scientific metaphors do: it made it easier to think, and therefore easier to discover.

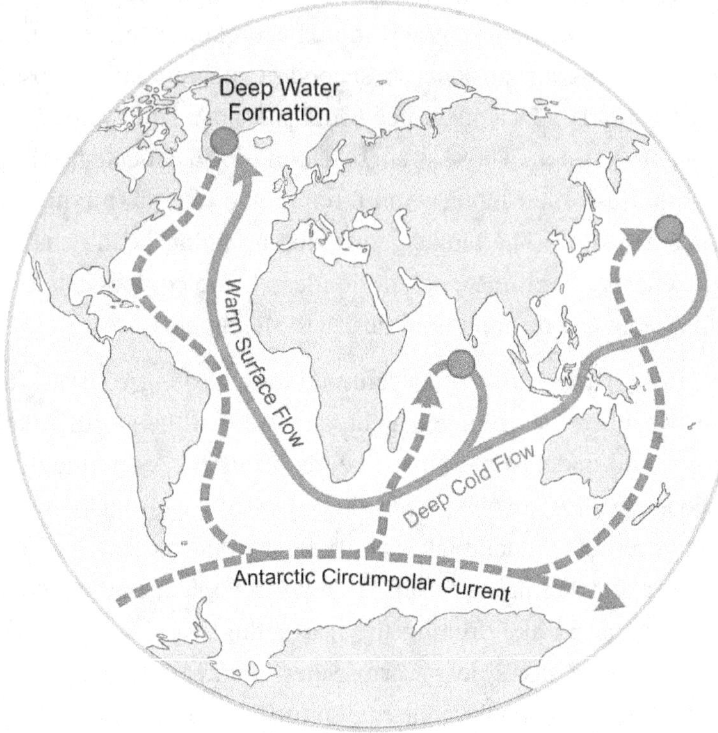

For a metaphor built from such broad strokes, the conveyor has proven remarkably durable. Not because it is perfectly accurate, but because it directs attention to the ocean's most consequential trick: taking surface water that has recently touched the sky and, in a few key places, transforming it into deep water that will not see daylight again for centuries. When the North Atlantic performs that transformation, it does more than stir the ocean. It helps set the background conditions for climate, for the ocean's ability to store carbon and heat, and for the chemical breath of the deep sea.

1.2 The Global Radiator

A useful way to think about the ocean is as a heat engine that never turns off. Sunlight pours energy into the tropics year after year, far more than the tropics can radiate back to space. High latitudes, receiving weaker sunlight and losing heat efficiently to cold, dry air, run an energy deficit. If the planet did not move heat poleward, the equator would grow steadily hotter and the poles steadily colder until the temperature contrast became absurd. The atmosphere does much of the work, flinging warm air and water vapour toward the poles in storms. But the ocean is the planet's slow, heavy radiator: it stores enormous amounts of heat and releases it where the air is most eager to take it away.

The ocean's role is easy to underestimate because most of us experience it as weather at the surface. We see the wind ruffle waves and we hear about famous currents like the Gulf Stream. Yet the most consequential feature is not the speed of a particular current; it is the *quantity of heat* transported, which depends on how much water moves and how different its temperature is from the surroundings. Even a slow flow can matter if it involves a huge volume.

The basic arithmetic is plain enough to write on a napkin. The heat carried by a moving body of water can be approximated as

$$H \approx \rho\, c_p\, V\, \Delta T,$$

where ρ is the density of seawater, c_p is its heat capacity, V is the volume transport (how much water flows per second), and ΔT is the temperature contrast between that flow and what would otherwise occupy its place. Seawater has a large heat capacity, which is why coastal cities tend to have gentler winters and summers than inland ones: the ocean can absorb and release heat without changing temperature as quickly as land.

This is where a common confusion creeps in. The Gulf Stream, pressed against the American coast, is a swift, narrow river in the sea. The Atlantic Meridional Overturning Circulation (AMOC), by contrast, is not a single current you can point to on a map. It is a basin-scale pattern: a northward flow of relatively warm, upper-ocean water; a transformation to dense water in the high-latitude North Atlantic; and a return flow of colder deep water southward. The Gulf Stream is part of the upper branch that helps deliver warm water poleward, but it is not identical to the overturning itself. One is a fast boundary current shaped heavily by winds and Earth's rotation; the other is a three-dimensional balance of flows and density changes that reaches from surface to abyss.

Why does this distinction matter? Because public discussions often slide from one to the other. People hear that the Gulf Stream keeps Europe mild and worry that it will stop like an engine stalling. In reality, wind-driven currents respond quickly to changes in winds, while the overturning involves dense-water formation and deep flows that evolve over longer timescales. The climate effect people care about is not tied to a single coastal current; it is tied to the ocean's ability to move heat northward across the Atlantic as a whole, and to release that heat to the atmosphere in the right places and seasons.

Nowhere is the ocean's "radiator" role clearer than the North Atlantic in winter. Cold continental air pours off North America and Greenland. It sweeps over relatively warm water and, in doing so, it drinks in heat and moisture. That exchange powers some of the most energetic storms on Earth. If you want a concrete sense of magnitude, the wintertime heat loss from the ocean in the subpolar North Atlantic can reach hundreds of watts per square meter during outbreaks. Spread across a large area, those fluxes add up quickly. The ocean is, in effect, paying out heat that it earned in the tropics.

This heat release is not a gentle warming blanket. It helps steer the atmosphere. Warmer ocean surfaces feed more evaporation, which alters humidity and the potential energy available to storms. The temperature contrast between warm water and cold air changes the stability of the lower atmosphere. In this way the ocean does not merely follow atmospheric winds; it shapes the patterns of weather that generate those winds. "Climate" can feel like a distant abstraction, but in the North Atlantic the connection runs through familiar experiences: storm tracks, winter rain, the frequency of cold snaps, and the chance of snow turning to sleet.

Europe's mild winters at its latitude are an everyday example of this redistribution. Cities such as London or Amsterdam sit at latitudes comparable to parts of Canada that endure much harsher winters, yet Europe's climate is softened by westerly winds that carry marine air inland. Those winds are part of the atmosphere's transport, but they are warmed by the ocean beneath them. The North Atlantic's heat release changes the temperature of the air that the prevailing winds deliver to Europe. Without the ocean's poleward heat transport and the resulting heat loss to the atmosphere, the prevailing westerlies would arrive colder and drier. The effect is not uniform across Europe, and it is not a single-cause story, but the ocean's contribution is unmistakable when you stand on a windy winter coast and feel air that is cold but not arctic.

There is another reason the radiator metaphor works: it invites comparison with the atmosphere. The Earth system divides the job of moving energy poleward. Near the equator, the ocean can carry a large share because sunlight warms the surface ocean efficiently and winds drive strong upper-ocean flows. Farther poleward, the atmosphere tends to dominate, especially in the mid-latitudes where storm systems are the main transport mechanism. The balance is not fixed; it depends on season, hemisphere, and the evolving climate state. But the overall picture is robust: the ocean and atmosphere share the labour, and the ocean is particularly important in the subtropics where it can carry heat poleward with relatively little loss to space.

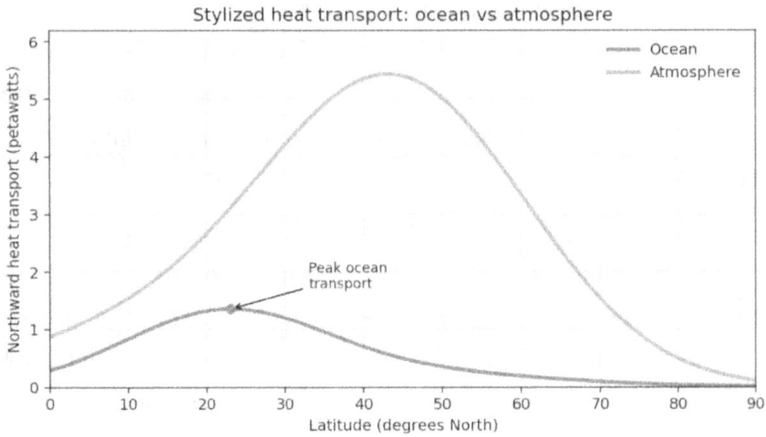

If the ocean is a radiator, the AMOC is often described as its pump in the Atlantic. That language carries a tempting implication: weaken the pump and you automatically weaken poleward heat transport. The truth is subtler, and appreciating the subtlety is a mark of understanding rather than a reason for despair.

One complication is that much of the ocean's poleward heat transport is carried in the upper ocean, in flows shaped strongly by winds and by the rotation of the Earth. The deep return flow matters for the global balance of mass and for the long-term storage of heat and carbon, but changes in the deep limb do not translate one-for-one into changes in total heat transport. Work such as Boccaletti and Ferrari's (2005) analysis of the "vertical structure" of heat transport helped crystallize this point: the ocean can move heat poleward mostly in shallow layers, while the deep ocean can be crucial for other reasons—ventilation, stratification, and the capacity to hide heat from the atmosphere for decades to centuries.

A second complication is that the AMOC is not directly observed over centuries in the way surface temperatures are. For a long time, it was inferred indirectly from hydrographic sections and from tracers, stitched together with physical constraints. Ganachaud and Wunsch (2000), for example, used hydrographic data in an inverse framework to estimate transports and heat fluxes, and the subsequent correction published in 2001 is a reminder that even careful synthesis carries uncertainty. The ocean is hard to measure because it is huge, variable, and expensive to sample.

This is why the start of continuous AMOC monitoring at 26°N in the Atlantic was such a milestone. The RAPID array, operating since 2004, transformed a subject that had often relied on snapshots into one with a daily time series. It also brought a surprise that is healthy for any science: the AMOC swings around a lot from year to year. A few years into the record, Smeed and colleagues (2014) reported a decline from 2004 to 2012, with the later years weaker than the early years in their analysis. That finding became widely discussed, sometimes with more certainty than it deserved, because a decade is long enough to worry the public but short enough to confuse trend with variability.

Then came the inevitable next step: attempts to reconstruct longer histories using other observations and statistical methods. Worthington and colleagues (2021) produced a 30-year reconstruction that did not show a simple monotonic decline, underscoring how sensitive conclusions can be to the method and period chosen. The IPCC's Sixth Assessment Report (2021) took a cautious stance that reflects this observational reality: it assessed that the AMOC is *very likely* to decline over the 21st century under all emissions scenarios, while expressing medium confidence that an abrupt collapse before 2100 is unlikely. Translated into ordinary language, that means the odds favour weakening, the consequences of weakening could be serious, and the most dramatic outcome is not the central expectation for this century.

These nuances matter because the ocean's radiator function is not a switch that flips from "on" to "off." It is more like a set of interacting valves that can shift how and where heat is released. A modest weakening of overturning could still have outsized effects on regional climate if it changes the pattern of sea surface temperatures in the North Atlantic. It could shift rainfall belts, alter the frequency of marine heatwaves, and influence the path and intensity of storms. In other words, even if the *total* heat transport changes less than a dramatic metaphor might suggest, the *distribution* of heat—where the ocean releases it to the air—can change in ways people feel.

A historical anecdote captures why this feels personal. During World War II, the North Atlantic was not merely a battleground for ships; it was a battleground for forecasts. The difference between ice-free waters and sea ice could determine routes. The difference between a storm track shifted a few hundred kilometres north or south could determine whether aircraft could fly, whether convoys could be protected, whether harbours could be used. Military meteorology pushed governments to invest in observing networks and forecasting techniques, and the North Atlantic's ferocious coupling of ocean and atmosphere was central. The war did not teach scientists the physics of heat transport from scratch, but it sharpened an already emerging truth: in this region, the ocean's surface is not passive scenery beneath the sky; it is an active participant in shaping the sky's behaviour.

The ocean radiator also has a less visible consequence that is becoming more important as the planet warms: the ocean has absorbed the overwhelming majority of the extra heat trapped by greenhouse gases. That fact can sound like good news—as if the ocean is kindly buffering us—and in the short term it is. Without the ocean's heat uptake, the atmosphere would have warmed far more rapidly. But heat stored in the ocean does not vanish. It can resurface later, and it can reorganize the stratification that controls mixing and the formation of dense waters. A warmer upper ocean can become more layered, with lighter warm water sitting atop colder water, making it harder for winter cooling to trigger deep mixing in places where it once did. The radiator can, in this sense, begin to insulate itself.

There is a final, human-scale way to appreciate the radiator metaphor. Think of a coastal city in winter—Boston, Dublin, Bergen. The ocean nearby is not warm in any absolute sense, but compared with the air during a cold outbreak it can be a reservoir of energy. When the wind blows from land to sea, the ocean gives up heat; when the wind shifts, that marine air is delivered back to land. The discomfort of a wet wind, the ease with which snow turns to rain near the coast, the way a mild spell arrives with a particular smell of salt and thaw—these are sensory traces of a planetary energy balance being negotiated in real time.

A planet with oceans is not just a planet with water. It is a planet with a moving, heat-storing medium that connects the bright surplus of the tropics to the dark deficit of the poles, and that releases energy in bursts and gradients that the atmosphere turns into weather. The radiators of human houses are tame by comparison: they sit still, they warm one room, and they do not decide where the storms will go.

1.3 Architecture of the Abyss

A map of the ocean, even a beautiful one, is a kind of lie. It flattens what matters most about seawater: that it is arranged in layers, and that the layers move differently. The Atlantic has a surface that can change in a week and a deep interior that remembers winters from centuries ago. To understand why a circulation can be strong or fragile, why it can ventilate the deep sea or starve it of oxygen, it helps to stop thinking of currents as lines on a map and start thinking of the ocean as a building with floors.

Stand on the deck of a ship in the North Atlantic and look out. The surface looks restless, textured by wind, streaked with foam. But the surface is a thin skin. Beneath it, the upper few hundred meters form a layer that is mixed vigorously by waves and storms, and warmed and cooled by the air above. Oceanographers call this the *mixed layer*. In winter it can deepen dramatically as cold winds and storms stir the water column; in summer it can become shallow as the sun warms the surface and makes it lighter, creating a stable lid. This seasonal breathing matters because the deep ocean cannot easily exchange heat and gases with the atmosphere unless winter mixing opens a pathway.

Below the mixed layer lies the *thermocline*, a region where temperature drops rapidly with depth. In the subtropical Atlantic, a diver descending through the upper ocean would feel the water cool quickly; a sensor lowered from a ship would trace a sharp bend in the temperature profile. The thermocline is not only about temperature. It is a density gradient, a barrier that resists vertical motion. Warm, light water sits above colder, denser water, and the stratification acts like a spring: push a parcel down and it wants to rise back up; pull it up and it wants to sink. Stratification is the ocean's architecture. It determines how easily surface changes can penetrate downward, and how easily deep water can return to the surface.

In the Atlantic, the story becomes more dramatic below roughly a kilometer. Here the ocean is cold, dark, and comparatively uniform in temperature, yet it is not uniform in origin. The abyssal Atlantic is filled primarily by two great water masses that approach each other from opposite ends: North Atlantic Deep Water (NADW) spreading southward from the subpolar North Atlantic, and Antarctic Bottom

Water (AABW) spreading northward from the Southern Ocean along the seafloor. Between them lies a zone of mixing and transformation, where properties blend and where the deep limb of the global overturning threads its way through gaps in underwater topography.

These names can sound like labels in a museum, but they refer to something tangible: a parcel of seawater formed under specific conditions, carrying a recognizable signature. NADW is generally colder and fresher than the subtropical surface but warmer and less dense than the very coldest bottom water. It is relatively rich in oxygen when newly formed, because it has been in contact with the atmosphere recently. AABW is colder and denser still, created near Antarctica where intense cooling and the formation of sea ice concentrate salt and drive water to sink to the abyss. It creeps northward along the bottom of the Atlantic like a low fog that cannot lift, because nothing above it is heavy enough to replace it.

To picture the vertical structure of the Atlantic, imagine taking a knife and slicing the ocean from north to south, from Greenland down past the equator and on toward the Antarctic margin. At the top, warm surface water flows northward, especially in the western part of the basin, feeding the regions where the ocean loses heat most aggressively to the winter atmosphere. Beneath that, between about 1000 and 3000 meters, a broad wedge of NADW flows south. Underneath, hugging the seafloor, AABW pushes north. The layers are not cleanly separated; they interpenetrate and mix, and their thickness varies with location. But the overall arrangement is robust enough that oceanographers can recognize it repeatedly in hydrographic sections.

The question that makes the architecture feel alive is this: how does a parcel of water move from the surface to a depth of two or three kilometers? Not by slowly sinking everywhere. The ocean is too strongly stratified for that. Deep waters form in a few special regions where the usual barriers weaken.

The North Atlantic is one of those regions. In parts of the Labrador Sea and the Nordic Seas, winter cooling can be severe enough to deepen the mixed layer to great depths, sometimes more than a kilometer. When storms remove heat faster than the ocean can replace it, the surface becomes dense and convective overturning can occur: dense water sinks and is replaced by slightly lighter water from below, which is then cooled and sinks in turn. This is deep convection, and it is one of the ways the upper ocean communicates with the abyss.

Yet deep convection is only part of NADW's origin story. A large fraction of the dense waters that feed the deep Atlantic pass through narrow, dramatic choke points between Greenland, Iceland, and Scotland. These are the overflows across the Greenland—Scotland Ridge. In the Nordic Seas, water can become dense through cooling and changes in salinity, and then it spills southward through passages such as the Denmark Strait and the Faroe Bank Channel. The spilling water is heavier than what lies downstream, so it descends the continental slope as an underwater waterfall, entraining surrounding water and transforming as it goes. The image is not poetic; it is literal. Oceanographers have measured these overflows and watched their properties change from decade to decade.

The Faroe Bank Channel overflow is a good example because it concentrates so much significance into a narrow place. A passage only tens of kilometers wide helps set the properties of a water mass that will fill vast volumes of the Atlantic at depth. This is one of the uncomfortable lessons of abyssal architecture: global-scale patterns can depend on the details of localized processes. A small change in the density of overflow waters, or in how much ambient water they entrain as they descend, can alter the downstream characteristics of NADW. In climate models, representing these overflows is challenging because they occur on scales smaller than the model grid, and because their dynamics depend on fine structure and topography. The IPCC has repeatedly noted that such process representation is a major source of uncertainty in projecting the magnitude of future AMOC change, even when the sign of weakening is robust.

Once NADW is formed and begins its southward spread, what keeps the circulation going? A tempting answer is "it sinks in the North and rises elsewhere," but that merely restates the puzzle. Water cannot rise through a stable density gradient without help. The help comes in two main forms: the winds, especially in the Southern Ocean, and mixing in the ocean interior.

The Southern Ocean is often described as the place where the deep ocean breathes. Strong westerly winds circle Antarctica, driving surface waters northward in an average sense through Ekman transport. Because the Antarctic Circumpolar Current has no continental barrier, these winds can induce a large-scale upwelling of deep waters toward the surface. Eddies complicate and partially counteract this tendency, but the net effect is that the Southern Ocean is a crucial gateway where deep waters can be brought upward and transformed. Lynne Talley (2013) emphasized in her synthesis of global overturning that the closure of the circulation is not a simple loop drawn on a map, but a set of intertwined pathways involving Southern Ocean dynamics, interior mixing, and water-mass transformations in the Indian and Pacific as well as the Atlantic.

Mixing, meanwhile, is the less glamorous but indispensable actor. In a perfectly still, stratified ocean with no mixing, the deep ocean would remain isolated, and overturning would eventually stall. Real oceans mix because they are never perfectly still. Tides slosh against rough topography, generating internal waves that propagate and break, stirring water across density surfaces. Winds inject mechanical energy into the surface, some of which cascades down into the interior. Even tiny turbulent motions, integrated over the vast volume of the ocean and over long times, can matter profoundly. Carl Wunsch and collaborators, among others, made a persuasive energetics argument: maintaining the observed stratification and overturning requires a supply of mechanical energy, and tides and winds are primary contributors. This shifts the mental picture again. The deep circulation is not powered solely by cooling in the North Atlantic; it is enabled by a global budget of energy that allows water to change its density class through mixing.

To feel why this matters, return to the building metaphor. Stratification is like a stack of floors with locked doors. Surface forcing—the push and pull of heat and freshwater at the surface—tries to move water between floors by changing its density. But the locks are tight. Mixing is the set of keys, or perhaps the wear and tear on hinges, that allows small exchanges between floors over time. Without mixing, the floors become isolated apartments. With it, there is slow but persistent exchange. The ocean's memory and its responsiveness are both controlled by how permeable these density "doors" are.

Hydrographic surveys carried out during programs such as WOCE (the World Ocean Circulation Experiment) provided the kind of global snapshots needed to test these ideas quantitatively. When scientists use inverse methods on such data, they essentially ask: given the observed temperature, salinity, and tracer distributions, and given the constraints of mass and energy conservation, what circulation and mixing rates are consistent with reality? Ganachaud and Wunsch (2000) produced influential estimates of global transports and mixing using such techniques, while also making clear the uncertainties involved. These are not merely technical exercises. They are attempts to close the global budget: to make sure that the imagined pathways of deep water formation and return flow add up to something physically plausible.

The Atlantic's vertical architecture also helps explain why the ocean can store heat and carbon for so long. When surface waters sink and become part of NADW, they carry with them dissolved gases and chemical properties set by their last contact with the atmosphere. Oxygen is particularly vivid. Newly formed deep waters can be oxygen-rich, supporting deep-sea ecosystems and controlling which chemical reactions dominate in sediments and water columns. If ventilation slows, oxygen can decline and the deep ocean's chemistry can shift. Carbon follows a different but connected logic. The ocean takes up carbon dioxide at the surface; biological processes and chemistry convert part of it into forms that can be transported downward; and deep waters can store it away from the atmosphere. The deep ocean is therefore both a buffer and a delayed messenger. What enters the abyss does not remain there forever, but it can be kept out of the atmosphere long enough to shape climate trajectories on human timescales.

A concrete way to link this to lived experience is to consider sea level along the Atlantic coasts. The AMOC and the density structure of the Atlantic influence sea surface height patterns. When the overturning weakens, it can alter the gradient of sea level across the ocean, with potential consequences for coastal sea level on the western Atlantic margins. To a person living near the shoreline, sea level rise can feel like a simple, uniform upward creep. In reality it is modulated by regional dynamics, including changes in currents and the density of seawater. The abyss, though far below sight, can thus contribute to how quickly water climbs a seawall in a storm surge.

Fragility, in this architecture, does not mean that the circulation is about to fail tomorrow. It means that key processes are localized and therefore potentially sensitive. The overflows across the Greenland—Scotland Ridge are one example: narrow passages with outsized consequences. Deep convection regions are another: they depend on weather patterns, sea ice, and the preconditioning of the water column, all of which can change. The Southern Ocean's upwelling and mixing pathways are yet another: they depend on winds and on the complex dance between eddies and mean flow.

This is why the vertical structure cannot be treated as a static diagram. It is a dynamic balance between buoyancy forcing (the push and pull of heat and freshwater at the surface), mechanical forcing (winds and tides supplying energy for mixing), and geometry (the shape of basins and ridges constraining where dense water can spill and where deep currents can pass). The architecture is not an accidental feature; it is the outcome of these forces negotiating with one another over millennia.

There is a quiet drama in realizing how much of this takes place beyond human senses. The deep Atlantic is cold, dark, and under immense pressure, yet it is not inert. It flows, mixes, ventilates, and stores. It carries the imprint of winters in the Labrador Sea and of gales over the Southern Ocean. It carries, increasingly, the imprint of human industry through heat and carbon that have found their way below the thermocline. And it depends, in surprising ways, on narrow underwater passages where dense water spills like a hidden cataract into the abyss.

Chapter 2

The Weight of Water

To the naked eye, the ocean appears as a singular, uniform expanse, yet beneath the waves lies a structured universe defined by the subtle interplay of heat and salt. This chapter peels back the surface to reveal how microscopic differences in density weigh entire oceans, driving the relentless vertical engine that powers the global climate. We examine the invisible forces that compel water to sink, the planetary rotation that steers it, and the turbulent energy required to bring it back to the light.

2.1 The Geometry of Heaviness

The ocean's grandest movements begin with a quiet accounting problem: how heavy is this parcel of seawater compared with the one beside it? The answer is not written on the surface. It is encoded in two ingredients that feel almost like opposites—heat and salt—and in a third that is easy to forget until you dive, namely pressure. Put them together and you get what oceanographers call the *equation of state*: a rule that turns temperature and salinity (and pressure) into density. Density is simply mass per unit volume, but in the ocean it acts like destiny. A slightly denser parcel tends to sink; a slightly lighter one tends to rise. Over the basin scale, the ocean's circulation is a choreography of these minute buoyancy differences, accumulated into currents large enough to move climate.

Yet the equation of state is not a straight line you can sketch on the back of a napkin. If it were, the ocean would be easier to predict and far less surprising. Warm water is generally lighter than cold water, and fresh water is generally lighter than salty water, but the way those effects combine is stubbornly nonlinear. Sometimes two very different waters can weigh the same. Sometimes mixing two waters produces something denser than either parent. And near freezing, seawater behaves with a particular ruthlessness: small changes in salinity can outweigh large changes in temperature, and the densest waters on Earth are manufactured in only a few places where winter, wind, and ice conspire to concentrate salt and strip away heat.

A useful way to picture this without getting lost in algebra is to imagine density as a landscape draped over a map whose axes are temperature and salinity. Pick a pressure, say the sea surface for simplicity, and ask: for every combination of temperature and salinity, what density would seawater have? The landscape is not a plane. It curves. It folds gently in some regions and steepens in others. On that landscape, there are contour lines where density is constant; oceanographers call these *isopycnals*. They are to seawater what contour lines on a topographic map are to hills. Where the contours bunch up, density changes rapidly with small changes in temperature or salinity; where they spread out, density is relatively insensitive.

The ocean lives on those contours. A parcel of water can slide along an isopycnal with little work, much as a bicycle rolls easily along a level road. Moving across isopycnals is like pedaling uphill: it requires energy, because it means making water heavier or lighter by changing its heat content or its salt content (or compressing it). The upper ocean is full of processes that push water across these density contours—sunlight warming the surface, evaporation leaving salt behind, rain diluting, sea ice rejecting brine, and turbulent mixing blending neighbors. The deep ocean, by contrast, is remarkably stratified: layered by density, arranged so that light water sits atop heavy water. That arrangement is stable in the same sense that oil floats on water; it resists being overturned.

The word *stratification* comes from the idea of layers, and it is the ocean's default state. You can think of the ocean as a stack of slightly different fluids, each with its own density, sliding past one another far more easily than they exchange places. This is why the ocean can store heat for decades to centuries: much of the heat absorbed near the surface stays trapped above denser layers. It is also why the deep ocean can remain isolated from the atmosphere long enough to become cold, oxygen-poor, or carbon-rich, depending on where it was last in contact with the surface.

At the center of all this is a deceptively simple question: which matters more, temperature or salinity? The only honest answer is: it depends on where you are on the temperature—salinity map. In temperate waters, cooling a parcel by a degree Celsius can increase its density roughly as much as adding a few tenths of a unit of salinity (measured in practical terms as something like grams of salt per kilogram of seawater, though oceanographers handle salinity with careful definitions). But that trade-off is not constant. Near the freezing point, the leverage of temperature diminishes: you cannot cool below freezing without making ice, so temperature runs out of room. Salinity does not. A small increase in saltiness at very low temperatures can have an outsized effect on density. This is one reason polar oceans are such potent "factories" of dense water.

There is also a pressure trick. Seawater is compressible: squeeze it with the weight of kilometers of water overhead and it becomes denser, even if its temperature and salinity stay fixed. The compressibility is not huge, but over ocean depths it matters. Pressure also changes the way temperature affects density. In the deep ocean, warming a parcel does not make it buoyant in quite the same way as it would near the surface. Oceanographers often speak of the *thermal expansion coefficient* and the *haline contraction coefficient*—measures of how strongly density responds to temperature and salinity—and both vary with temperature, salinity, and pressure. The coefficients are part of the reason the isopycnals in a temperature—salinity diagram curve rather than running as straight lines.

The temperature—salinity (T—S) diagram is one of oceanography's most elegant inventions because it compresses a sprawling three-dimensional ocean into a two-dimensional fingerprint. Each measurement of seawater can be plotted as a point with temperature on one axis and salinity on the other. Collections of points trace out curves that correspond to particular water masses: Antarctic waters, subtropical thermocline waters, intermediate waters formed in specific seas. The power comes from adding the isopycnals: curved lines of constant density laid over the scatter of points. With that overlay, you can see at a glance which parcels are heavier, which are lighter, and how mixing might transform them.

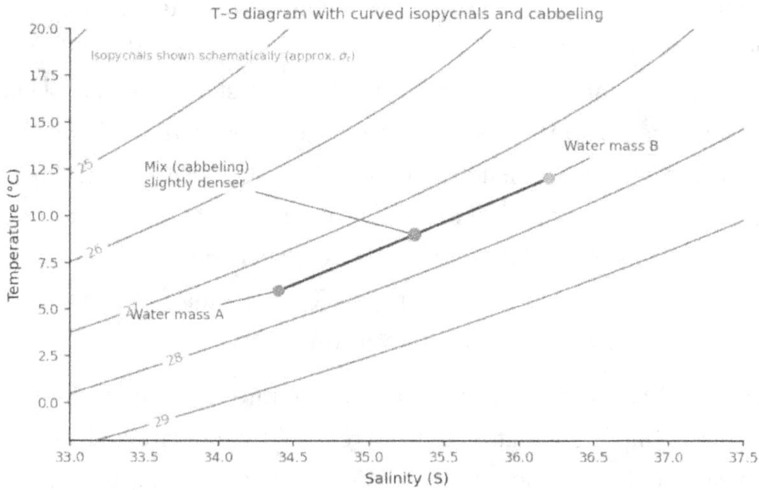

A T—S diagram also reveals one of seawater's strangest quirks: *cabbeling*. The name is old-fashioned, borrowed from sailors' language, but the effect is sharp. Suppose you have two parcels of seawater that happen to share the same density, even though one is warmer and saltier while the other is cooler and fresher. If seawater density depended linearly on temperature and salinity, then mixing them would simply produce water of the same density—an unremarkable average. But because the equation of state is curved, the average can bulge toward higher density. The mixture can be denser than either parent, just heavy enough to sink.

This is not a mathematical curiosity; it is a real pathway by which the ocean can create dense water without extreme cooling or extreme salting. Consider a boundary where warm, salty water meets cold, fresh water: a front, or the edge of a current, or the mixing zone around sea ice. Turbulence, eddies, and stirring smear the boundary. Where the two waters blend, the nonlinear equation of state can convert mixing into sinking. The ocean, in effect, gets a little extra gravity for free.

There is a companion phenomenon with a more threatening reputation: *double diffusion.* Heat and salt do not move through water at the same molecular rate; heat diffuses faster than salt. Under certain stratifications—warm, salty water over cold, fresh water, or the reverse—this mismatch can drive slow, staircase-like mixing patterns, visible in some regions as layered steps in temperature and salinity profiles. It is a reminder that even when water looks still, it can be rearranging itself by microphysical rules that have basin-scale consequences when integrated over time.

The practical business of turning temperature and salinity into density is so important that it has been repeatedly standardized, debated, revised, and made more precise as measurements improved. A concrete historical marker comes from the era when computers were becoming part of field oceanography. In 1983, a UNESCO technical report by Nick Fofonoff and Robert Millard gathered algorithms for seawater properties that underpinned what became known as EOS-80, paired with the Practical Salinity Scale (PSS-78). This was not glamorous work. It was the infrastructure of a science that wanted different ships, labs, and nations to compute density the same way.

As instrumentation sharpened and climate questions demanded tighter accuracy, that infrastructure was rebuilt. A newer international framework, TEOS-10, shifts the underlying thermodynamics to a Gibbs free energy formulation and replaces some familiar variables with more physically consistent ones, such as *Absolute Salinity* and *Conservative Temperature*. The motivation is not pedantry. Salinity is not a single substance you can bottle; it is a practical measure inferred from conductivity, and the ocean's dissolved ions are not perfectly uniform from place to place. Temperature, similarly, is not just a number on a thermometer when pressure changes and heat capacity matters. These refinements matter most when one is balancing budgets of heat and freshwater over entire basins, or when small density differences determine whether water sinks into the abyss or stays near the surface.

Still, for intuition, it helps to keep hold of a few physical anchors. Start with freshwater. Pure water has its own famous nonlinearity: it is densest at about 4°C, which is why lakes can freeze on top while remaining liquid below. Seawater behaves differently because dissolved salts disrupt that peculiarity and push the freezing point downward. For typical ocean salinity, seawater freezes near -1.8°C, and over most oceanic ranges it becomes steadily denser as it cools. This makes the ocean more straightforward than lakes in one sense: cold water tends to sink. But salinity reintroduces complication. The saltier the water, the lower its freezing point and the higher its density. Those two facts intertwine in polar regions in a way that makes sea ice a powerful density engine.

When seawater freezes, most of the salt does not fit into the ice crystal lattice. The forming ice rejects salt, leaving behind *brine*: a cold, very salty liquid that drains downward like syrup. If you have ever watched salt melt ice on a winter sidewalk, you have seen the same physics in miniature: salt changes phase boundaries and produces dense, cold fluid. In the ocean, brine rejection makes the water beneath ice both colder and saltier—a double push toward higher density. In certain places, with the right winds and geography, this process creates the densest waters that fill the deep basins.

A concrete example is the Weddell Sea, a vast embayment of the Southern Ocean bordered by Antarctica. In winter, fierce katabatic winds pour off the ice sheet, opening and maintaining polynyas—stretches of open water or thin ice amid the pack. These openings lose heat rapidly to the cold air. New ice forms, rejects brine, and the surrounding seawater grows denser until it sinks and spreads along the seafloor as Antarctic Bottom Water. The story is not simply "cold sinks": it is the combined geometry of cooling and salting, operating in a narrow part of the T—S space where density increases steeply.

A similar, though not identical, factory exists in the North Atlantic. There, surface waters carried northward by currents release heat to the atmosphere, becoming colder and denser. In regions like the Labrador Sea and the Nordic Seas, winter storms can deepen the surface mixed layer dramatically, stirring cold conditions downward and allowing the ocean to ventilate intermediate and deep layers. Again, salinity is decisive. Freshwater input from precipitation, rivers, and melting ice can lighten surface waters enough to cap the ocean, preventing deep convection even under intense cooling. This sensitivity is one reason the North Atlantic occupies such an anxious place in climate discussions: the density structure depends on a tug-of-war between heat loss (making water heavier) and freshwater addition (making it lighter).

The nonlinearity also means that the same surface cooling can have different outcomes depending on what water you start with. Cooling a warm, relatively fresh surface layer might make it denser but still not dense enough to sink below a salty layer underneath. Cooling a salty layer by the same amount could trigger sinking. In other words, the ocean's response to atmospheric forcing is conditional. The atmosphere can push; the ocean decides whether it yields.

It helps to make the relationship more tangible with a thought experiment at the scale of a bucket. Imagine two buckets of seawater: Bucket A is cool and moderately fresh, Bucket B is warm and quite salty. Measured carefully, their densities could be nearly identical. If you pour them into a third bucket and stir, you might expect the result to sit neutrally. But because the density contours curve, the mixed water can end up slightly denser and therefore slightly more eager to sink. Now enlarge the buckets to the size of water masses and the stirring to the size of ocean eddies. Cabbeling becomes one of the subtle ways the ocean turns horizontal contrasts into vertical motion.

The geometry matters in another, even more consequential way. Density does not just decide who sinks; it sets the pressure field. Heavier water columns press more strongly on the water beneath them. Differences in density from place to place create differences in sea surface height and internal pressure surfaces. Those pressure gradients are the immediate drivers of large-scale currents, though on a rotating Earth they rarely drive flow directly downhill in the way intuition suggests. The point here is simpler: the ocean's density structure is not a passive record of heating and freshening; it actively shapes the forces that move the ocean around.

Oceanographers often prefer to talk about *buoyancy* rather than density itself. Buoyancy is essentially "how much a parcel wants to rise or sink" compared with a reference, and it is the variable that enters the equations of motion neatly. But buoyancy inherits all the equation of state's quirks. If you treat temperature and salinity as independent sliders, buoyancy does not respond with equal smoothness everywhere. In some regimes, temperature dominates; in others, salinity does; and in still others, it is the cross-term—the way temperature and salinity modify each other's effects—that gives the ocean its unexpected behaviors.

Pressure's role deserves one more concrete image. Take a parcel of water at the surface and imagine moving it straight down without letting it exchange heat or salt with its surroundings. It gets compressed, and its temperature rises slightly due to adiabatic compression (much as air warms when compressed). The surrounding ocean also changes with depth. Whether the parcel is heavier or lighter than its surroundings at the new depth depends on how its properties compare along the correct thermodynamic path. This is why oceanographers use careful temperature definitions when making density comparisons across depth. The deep ocean is not simply "cold"; it is cold *at high pressure*, and that modifies the relation between temperature and density.

All of these subtleties funnel toward a practical outcome: the ocean tends to arrange itself into stable layers, but it has a few select pathways for punching through that stability and creating very dense water. Those pathways are geographically constrained, because they require the right combination of forcing and preexisting water properties. The ocean is not equally willing to make bottom water everywhere. It needs cold air, strong winds, ice processes, or some combination that can push surface waters across the tightest density contours. The densest waters form where the T—S geometry makes density hypersensitive to winter's work.

Once formed, these dense waters spread far from their source. They do not carry a label, but they do carry a signature: a particular blend of temperature and salinity that marks their origin. On a T—S diagram, they occupy distinct regions, and their gradual modification by mixing traces out lines and curves that can be read like a history of encounters. In that sense, the equation of state is not only a rule for calculating density. It is a language for telling the ocean's story: how heat and salt were rearranged, where water was last exposed to the atmosphere, and why a seemingly gentle act like mixing can sometimes create the heavy water that slips into the deep and stays there, pressing its weight on the rest of the circulation.

2.2 The Great Deflection

Stand on a windy headland and the ocean looks like a creature of instant moods: chop, whitecaps, a slick of foam racing downwind. It is tempting to imagine that wind simply pushes the sea along, the way a hand pushes a curtain. Near the surface, that picture is not entirely wrong. But it fails in a particularly revealing way: the sea does not obediently flow with the wind. It turns.

The turning is not a quirk of coastlines or currents. It is the signature of a rotating planet. Earth's spin quietly inserts itself into almost every large-scale movement of the ocean, and it does so with a force that feels like magic until you learn to recognize its rules. The result is a two-layer personality. The ocean's skin is fast, impulsive, and directly harassed by weather. The interior is slower and more constrained, moving along lines set by pressure and rotation in a state called *geostrophic balance*. The dividing line between them is not a solid boundary; it is a zone of fading turbulence and dwindling friction, where the ocean hands off control from the wind to the planet.

The stage was set by the density story already on the table. Differences in temperature and salinity create differences in density, and differences in density create pressure gradients: columns of lighter water sit a little higher, columns of heavier water sit a little lower. If Earth did not rotate, water would accelerate from high pressure toward low pressure until the pressure field flattened out. The ocean would be eager to "spill" like water on a tilted floor. That is the intuitive picture most people carry from everyday fluids.

But the open ocean is not a sink full of dishwater. At the scales of hundreds to thousands of kilometers, friction is weak enough that another balance becomes possible. A parcel of water that tries to rush downhill is deflected sideways by the Coriolis effect, an apparent force that arises in rotating systems. In the Northern Hemisphere, the deflection is to the right of the motion; in the Southern Hemisphere, to the left. The Coriolis effect does not create speed out of nothing. It redirects speed, like a patient hand turning the handlebar of a bicycle. If there is no motion, there is no Coriolis deflection. Yet because the ocean is constantly being nudged into motion by wind, tides, and pressure gradients, the deflection becomes a governing influence.

A concrete way to feel this without equations is to imagine a broad mound of water in the middle of the ocean, a gentle sea-surface hill only a meter or two high spread across a basin. Gravity wants the water to slide down the slope, away from the mound. If it does, it begins to move. As soon as it moves, the Coriolis effect bends the flow sideways. With enough scale and enough time, the flow can reach a steady compromise: gravity continues to pull water down the pressure slope, while the Coriolis effect continually turns it sideways. The water ends up flowing *along* the contours of pressure rather than straight down them. That compromise is geostrophic balance. It is the oceanic equivalent of air circulating around a high- or low-pressure system in the atmosphere.

The word “geostrophic” sounds forbidding, but the idea is simple: *geo* for Earth’s rotation, *strophic* for turning. In geostrophic balance, the sideways Coriolis deflection exactly matches the sideways push needed to keep water from sliding down the pressure gradient. The ocean does not lose the pressure difference; it converts it into a current that wraps around it.

This balance is what makes the interior ocean feel so different from the surface. Near the surface, the wind applies stress directly, and the ocean responds with an *Ekman layer*: a thin, frictional layer, typically tens of meters thick, in which the current rotates with depth and the net transport is at right angles to the wind. The classic result, worked out by Vagn Walfrid Ekman in 1905, came from a remarkably practical puzzle. Fridtjof Nansen had noticed during the Arctic drift of the *Fram* that sea ice did not move exactly with the wind. The ice drifted at an angle. Nansen asked his student Ekman to explain it, and Ekman found that a rotating Earth plus friction could produce a spiral of current direction with depth. The surface current is pulled by the wind but deflected by rotation; the layer beneath is pulled by friction from above and deflected as well; and so on, producing a graceful twist. Add up the transport over the whole layer and the net movement is 90 degrees to the wind: to the right in the Northern Hemisphere, to the left in the Southern.

That 90-degree rule is often presented as a slogan, but its consequences are anything but small. It means that winds do not merely push water; they *pile it up* and *pull it away* in specific patterns. Where winds drive surface waters together, water accumulates and the sea surface rises; where winds drive surface waters apart, water is removed and the sea surface falls. These subtle changes in sea level are not the stuff of beachcombing—they are typically measured in centimeters to tens of centimeters—but across an ocean basin they create pressure gradients strong enough to set up major currents.

Take the North Atlantic subtropical gyre as an example, a vast clockwise circulation that includes the Gulf Stream. Prevailing winds over the basin include the trade winds at lower latitudes and the westerlies at mid-latitudes. The pattern of wind stress causes Ekman transport that converges toward the gyre's center. Water accumulates there, forming a broad dome of elevated sea surface. Gravity wants to flatten the dome, but Earth's rotation turns the resulting pressure-driven flow into a current that circles around the mound. The gyre is maintained: wind creates the mound, geostrophy creates the swirl.

This is the ocean's great deflection in action: the wind does work at the surface, but the interior responds as a rotating, pressure-balanced fluid. The fast, scrappy surface and the steady interior are connected by the exchange of momentum and by vertical motions that, though often small in speed, are immense in their cumulative effect.

Those vertical motions are where the thermohaline story re-enters with force. Convergence and divergence in the Ekman layer cannot simply accumulate forever; water is nearly incompressible. When surface waters converge, they must go somewhere, and the only place to go is down. When surface waters diverge, water must come up from below. This is *Ekman pumping* (downward) and *Ekman suction* (upward), and it is a mechanism by which wind can reach into the ocean's interior without physically mixing it. The vertical velocities are typically tiny, on the order of meters per day or less, but spread over thousands of kilometers they reshape the distribution of heat and nutrients and influence where density surfaces rise and fall.

Upwelling regions are an everyday example with global consequences. Along the west coasts of continents in the subtropics—California, Peru, northwest Africa—winds often blow parallel to the coast. In the Northern Hemisphere, Ekman transport is to the right of the wind, which is offshore in these settings; surface water is pushed away from the coast, and colder, nutrient-rich water rises from below to replace it. The result is a narrow ribbon of cool sea surface temperatures visible from satellites and a fertile marine ecosystem that supports fisheries and seabirds. The surface looks lively and immediate, but the mechanism is a rotating, frictional boundary layer coupled to a slower interior that provides the replacement water. Even the fog rolling onto coastal cities has an oceanographic backstory written in Ekman transport and geostrophic structure.

Geostrophic balance does not mean the ocean is static or perfectly orderly. It means that, away from boundaries and storms, the dominant forces in the horizontal are pressure gradients and Coriolis deflection, and they nearly cancel. That "nearly" matters. The small leftovers—friction, eddy stirring, the time dependence of winds—allow the ocean to evolve. But the near-cancellation has a profound implication: the ocean interior can support currents that are swift yet require very little direct forcing, because they are not continually accelerating downhill. They are, in a sense, balanced against the planet.

This balance also provides one of oceanography's most powerful measurement techniques. If you know the ocean's density structure, you can infer its currents. Density determines pressure. Pressure gradients plus rotation give geostrophic flow. This is why a vertical profile of temperature and salinity, taken with a modern conductivity–temperature–depth instrument, can be turned into estimates of current shear—how the current changes with depth—even without directly measuring velocity. The ocean hides its motion in its stratification, and geostrophy provides the key.

A famous historical example of this logic becoming a working tool comes from the early twentieth century, when systematic ocean surveys were still rare and instruments were crude by modern standards. Researchers began to realize that mapping temperature and salinity was not just a descriptive exercise; it was a way to diagnose dynamics. Later, as satellite altimetry made it possible to measure sea surface height with centimeter precision, the same geostrophic principle could be applied from space: slopes in sea surface height correspond to horizontal pressure gradients at the surface, which correspond to surface geostrophic currents. In the broad strokes of modern ocean monitoring, the ocean's interior is still being inferred from its pressure field, and the pressure field is still being inferred from the ocean's density and height.

There is a deep reason geostrophy is so common: the ocean adjusts itself toward it. If the ocean is suddenly disturbed—say, a storm creates a patch of piled-up water and chaotic currents—the unbalanced pressure gradients generate motions that are quickly turned by rotation. The system radiates away some of its imbalance through waves and settles into a state where large-scale flows tend to run along pressure contours. This process, known in more technical language as *Rossby adjustment*, helps explain why the ocean can remain close to geostrophic balance much of the time even though it is constantly being perturbed. The ocean has a preferred posture, and rotation is the spine.

The preference has a speed limit attached. The effectiveness of Coriolis deflection depends on latitude: it vanishes at the equator and strengthens toward the poles. That is why the equatorial ocean behaves differently. Near the equator, winds can drive strong east–west currents and produce dramatic upwelling along the equatorial band because the usual sideways turning is weak and changes sign across the equator. The equatorial Pacific's upwelling and the associated sea surface temperature patterns are central to El Niño and La Niña, climate events that affect rainfall, drought, and storms across the globe. Here again, rotation is present, but its geometry changes; the great deflection becomes a delicate sign change that reorganizes how wind stress is translated into pressure and currents.

The distinction between the wind-driven surface layer and the geostrophic interior is not a clean separation of "fast" and "slow" everywhere, but it is a reliable contrast in how the ocean is controlled. Near the surface, friction matters because wind stress must be transmitted downward. Turbulence and waves make that layer messy, and the current can change quickly with weather. Deeper down, friction is weaker and the ocean tends to obey the large-scale balance between pressure gradients and Coriolis deflection. Variability still exists, but it is organized: eddies swirl like weather systems in the sea, and long waves propagate across basins with timescales of months to years.

Those eddies deserve a brief moment because they are where the two personalities meet. Mesoscale eddies—tens to hundreds of kilometers across—are the ocean's energetic middleweight. They are largely geostrophic, spinning around pressure anomalies, but they are born from instabilities that feed on the density structure and on currents that the wind helps maintain. Eddies stir heat and salt laterally along density surfaces much more efficiently than the ocean mixes them vertically across density surfaces. This is one of the reasons the ocean can transport climate-relevant properties over long distances without erasing its stratification. The interior remains layered, but within those layers the ocean is restless.

For someone thinking about climate, the crucial point is that the ocean's ability to store heat and carbon depends on the pathways by which surface waters communicate with the interior. The wind-driven layer can exchange heat with the atmosphere on timescales of days to seasons. The geostrophic interior can sequester anomalies for decades to centuries. The exchange between them is not just a matter of turbulence; it is also a matter of geometry on a rotating sphere. Wind patterns create convergences and divergences. Those create vertical motions that tilt density surfaces. Those tilts reshape pressure gradients. Pressure gradients drive geostrophic currents. The currents, in turn, redistribute temperature and salinity in ways that feed back on density. The ocean is constantly negotiating between being pushed around and being balanced.

A memorable, human-scale example of this negotiation is the Gulf Stream's relationship with the atmosphere. Mariners have known for centuries that a strong current flows northeast along the American coast and across the Atlantic, warming the air above it and steering weather. Benjamin Franklin, serving as a colonial postmaster, took an interest in the Gulf Stream because it affected mail delivery times. Working with his cousin Timothy Folger, a Nantucket whaling captain familiar with the current, Franklin helped produce one of the first maps of the Gulf Stream in the late eighteenth century. The map was practical, but its underlying story is physical: winds help maintain the subtropical gyre, the gyre piles up water, and geostrophic balance produces a swift western boundary current as the ocean returns poleward along the coast. The fast current that sailors felt under their hulls was, in a real sense, the ocean's way of balancing a basin-scale pressure field under rotation.

That western boundary intensification—the tendency for gyres to have strong, narrow currents on their western sides, like the Gulf Stream and the Kuroshio—is another place where Earth's rotation leaves a distinctive fingerprint. The Coriolis effect changes with latitude, and that variation helps squeeze the return flow into a jet along the western edge of the basin. The result is a current that can run at several knots, with sharp temperature contrasts across its edge, a boundary between water masses that can be traced on a T–S diagram and felt in the air as a sudden change in warmth and humidity.

Meanwhile, far below, the interior of the gyre moves slowly, almost imperceptibly to a ship. Yet it carries immense transport when integrated over depth and width. It is the ocean's quiet conveyor of properties, guided by the constraints of rotation and stratification. A glassy patch of sea at dawn can be above a current that is not calm at all, simply balanced: flowing steadily along a pressure contour that is itself maintained by the winds overhead and the density structure below.

The great deflection, then, is less a single effect than a disciplined habit the ocean acquires on a rotating world. When pushed, it does not simply yield in the direction of the push. It turns, organizes, and often responds at right angles. The wind writes its message on the surface, but rotation translates that message into a geometry of sea-surface slopes, pressure gradients, and currents that circle rather than spill. In the end, the ocean's motion becomes legible not as a chaotic response to weather, but as a persistent balance between the urge to level pressure differences and the planetary insistence on turning any attempt at straight-line motion into a curve.

2.3 Stirring the Abyss

The deep ocean has an awkward problem. Dense water can be made at the surface in a few hard, cold places—and once made, it sinks with an ease that looks almost effortless. Gravity does the work. But a global overturning circulation cannot be a one-way drain. If heavy water only fell and never returned, the abyss would steadily fill with the densest fluid available and the upper ocean would be left thin and depleted, a kind of oceanic bankruptcy. The real ocean avoids this fate. Water that sinks into the abyss is, somehow, lifted back toward the surface, lightened along the way, and returned to the parts of the world where it can again exchange heat and gases with the atmosphere.

That "somehow" is where the romance of the deep sea meets the bookkeeping of energy. Raising dense water requires work. Not metaphorical work, but mechanical energy, because lifting dense fluid increases gravitational potential energy. In the everyday world, the energy comes from engines and muscles. In the ocean, it comes from winds, tides, and the slow grinding of currents over rough seafloor topography. The abyss is stirred, not uniformly, but in patches and along boundaries, in bursts that are invisible to casual observation and yet decisive for the planet's climate.

The paradox is easy to state. The ocean is layered by density: light water above, heavy water below. That layering—*stratification*—is stable. If you try to mix the layers by pushing heavy water upward and light water downward, gravity pushes back. Stirring a stratified fluid is like stirring a cocktail that wants to unmix itself. The ocean does not like to move water across density surfaces, the way it happily slides water along them. In technical language, mixing *along* density layers is comparatively easy; mixing *across* them is difficult. The overturning circulation, however, demands exactly that difficult action. Bottom water must be transformed into lighter water, which means it must cross density layers—a process called *diapycnal mixing* ("across density").

For much of the twentieth century, the temptation was to imagine that the abyss is slowly and uniformly mixed by weak turbulence everywhere. This is the seductive simplicity of what Walter Munk famously called an "abyssal recipe" in 1966: if you assume an upward flow of deep water that is balanced by a uniform vertical mixing (often represented as a diffusivity), you can reproduce the broad shape of observed temperature profiles in the deep ocean. The model is not wrong in spirit; it captures the idea that a steady overturning requires an upward return. But it leaves out a question that later became unavoidable: where does the energy come from, and where does it actually get dissipated into turbulence?

The answer arrived not as a single revelation but as a tightening net of observations. One of the most vivid pieces of evidence came from tracer-release experiments, the oceanographic equivalent of putting a drop of dye in a river and watching it spread—except the "drop" is a carefully chosen chemical tracer released at depth and tracked over months and years. In a landmark experiment in the early 1990s, James Ledwell and Andrew Watson released a tracer near the main thermocline and measured how it spread. The rate at which it leaked vertically across density layers was startlingly slow. The open ocean interior, at least in that region, seemed too quiescent to provide the robust, uniform mixing that the simplest recipes had assumed.

This observation did not kill the idea of mixing; it forced a change in geography. If the interior is too calm, then mixing must be stronger elsewhere. And once you begin to ask where the ocean is rough, where energy is injected, and where waves might break, the map lights up in a very particular pattern. The abyss is not a gently simmering pot. It is more like a kitchen with a few powerful whisks:

narrow straits, steep ridges, continental slopes, and jagged seamount chains, places where tides and currents are forced to interact with topography.

To see why, it helps to remember that the ocean is full of waves that do not live on the surface. A stratified fluid supports *internal waves*, oscillations of dense and light layers sliding past one another. If you have ever seen cream swirl into coffee and form delicate layers, you have glimpsed how easily stratified fluids can support motion without mixing. In the ocean, internal waves can carry energy over long distances, threading through the interior without immediately dissipating. They are the ocean's hidden way of transporting mechanical energy from where it is generated to where it can finally be converted into turbulence.

Two major engines generate internal waves. The first is the tide. The Moon and Sun pull on the ocean, producing tidal currents that sweep back and forth across the seafloor. Where the bottom is smooth and flat, these currents are comparatively gentle. Where the bottom rises into ridges and sills, the tidal flow is forced up and over obstacles. In a stratified ocean, that up-and-over motion generates internal tides: internal waves with tidal frequencies that can propagate away from their birthplace, like sound from a struck bell.

The second engine is the wind, not by directly stirring the abyss, but by shaking the surface and producing near-inertial motions that radiate downward as internal waves. Storms can inject energy into the upper ocean at frequencies near the local inertial frequency (set by Earth's rotation), and that energy can leak into the interior as internal wave beams. The wind, which seems so superficial when you watch waves break, can thus feed the abyss indirectly through the internal wave field.

There is a third pathway that has become increasingly appreciated:

currents flowing over topography. Even without tides, steady or slowly varying flows can generate internal *lee waves* as stratified water is forced over seamounts and ridges, much as air flowing over a mountain range produces atmospheric mountain waves. These internal lee waves can break downstream, producing localized turbulence. In parts of the Southern Ocean, where strong currents sweep across rough seafloor features, lee-wave generation may be a major sink of kinetic energy.

All of these processes share a crucial step: wave breaking. Waves are not mixing by themselves; they are a way of storing and transporting energy. Mixing happens when waves become unstable and overturn, forming billows and turbulence that can irreversibly blend fluids of different densities. The deep ocean's "stirring" is therefore often a two-stage story: energy is injected at large scales (by tides, winds, or currents), converted into internal waves, and then dissipated into turbulence at smaller scales, often near boundaries or topographic features.

This perspective makes the deep ocean feel less like a uniform abyss and more like a landscape. Some places are loud, in a mechanical sense; others are quiet. Over the past couple of decades, compilations of microstructure measurements—instruments that can sense tiny fluctuations of velocity and temperature associated with turbulence—have reinforced this picture. Turbulent dissipation tends to be intensified near the bottom and near rough topography, not uniformly spread through the mid-water interior. The strongest mixing can occur above mid-ocean ridges, around seamount chains, and along continental slopes, where internal tides and waves are most likely to be generated and to break.

A concrete example is the Hawaiian Ridge. Strong tidal currents flow across this line of volcanic islands and submerged mountains. The ridge is a prolific generator of internal tides, and observations have shown energetic internal wave beams radiating away. Downstream, these waves can contribute to mixing at depth. The Hawaiian Ridge is not the only player—the Mid-Atlantic Ridge, the Luzon Strait, and many other features matter—but it provides a memorable illustration of how seafloor geometry can shape the ocean's energy pathways. The ocean's overturning is not driven only by what happens at the surface; it is also sculpted by the texture of the bottom.

It is worth pausing on an emotional truth hidden in these mechanisms: the abyss is hard to reach. For a long time, oceanographers inferred deep processes from sparse measurements and from the surface expression of currents. The shift toward thinking about internal waves and localized mixing was not merely conceptual; it was enabled by better instruments and patient campaigns that could measure tiny turbulent dissipation rates in hostile conditions. The deep ocean does not advertise its mixing. It whispers it, in fluctuations that can be smaller than a degree and swirls that die out in seconds. Yet those whispers integrated over the planet can determine how quickly the ocean ventilates and how long heat and carbon remain sequestered from the atmosphere.

To connect stirring to overturning, consider what it means to lift bottom water. One can imagine an "upwelling" of deep water, but it is not like a bubble rising. In the stratified ocean, water moves most naturally along density surfaces, which are often tilted and warped by currents and wind-driven pumping. For bottom water to become shallower in a sustained way, it must be transformed into a lighter density class. That transformation can happen in two broad ways. One is by adding heat, which tends to make water lighter. The other is by removing salt, which also tends to make water lighter (though in the deep ocean, salt changes are often mediated through mixing rather than direct freshwater flux). Both effectively require diapycnal mixing: exchanging properties across density layers.

Here the wind returns to the story in an unexpected role. Winds over the Southern Ocean, especially the strong westerlies that encircle Antarctica, drive an intense surface circulation and a pattern of upwelling that brings deep waters toward the surface. But the winds alone cannot complete the transformation. They can tilt density surfaces and pump water upward along them; they can expose deep water to the surface where it can be warmed and freshened. Yet the stratified interior still resists vertical exchange. The overturning circulation depends on a partnership: winds help pull deep waters up to where surface buoyancy forcing can lighten them, while internal-wave-driven mixing and boundary processes help reshape the density structure that allows this exchange to occur without the system grinding to a halt.

The energy constraint sits behind this partnership like a stern accountant. Carl Wunsch and Raffaele Ferrari emphasized in 2004 that maintaining the ocean's stratification and overturning requires a supply of mechanical energy. The ocean is constantly converting potential energy into kinetic energy and vice versa. Dense water formation increases potential energy by placing heavy water below light water in a stable configuration, but the act of forming dense water and sending it to depth also locks away the buoyancy anomalies that drive circulation. To keep the system moving, energy must be injected and dissipated. Winds inject energy into the upper ocean and into large-scale currents. Tides inject energy through their interaction with topography. Some of that energy is ultimately spent on mixing across density layers, the costly step that enables deep water to return.

There is a temptation to picture mixing as a kind of universal solvent: stir the ocean enough and everything blends. The abyss refuses that picture. Mixing is selective and uneven, and that unevenness can shape pathways of return flow. A parcel of bottom water might travel thousands of kilometers along a deep boundary current with little change, then pass over a rough ridge where internal waves and turbulence erode its density slightly, lifting it into a lighter class. It might then be swept along a different pathway, interacting with other waters, experiencing further gentle transformations. The return to the surface is not necessarily a single elevator ride; it can be a series of small promotions through the density hierarchy.

This view has practical consequences for how climate models represent the ocean. Models cannot resolve every breaking internal wave, and they cannot explicitly represent turbulence over every seamount. Instead, they use parameterizations: rules that represent the net effect of unresolved mixing. If mixing is geographically patchy, then representing it as uniform can produce the right answer for the wrong reasons, and the wrong answer when the climate changes. Changes in winds, sea ice, and stratification can alter the generation and propagation of internal waves, potentially shifting where mixing occurs. Changes in tidal energy itself are not expected to swing wildly on human timescales, but the pathways by which tidal energy becomes turbulence can be affected by stratification and circulation. In this sense, the geometry of the seafloor and the state of the climate become entangled through the ocean's energy cascade.

Even the language of "the abyss being stirred" can mislead, because much of the mixing that matters for overturning occurs not in the flat middle of basins but along their edges. Continental slopes, where the seafloor rises toward land, are places where currents rub against boundaries, internal tides interact with shelves, and turbulence can be enhanced. The ocean's margins are also where sediments, nutrients, and human impacts accumulate. The abyss may feel far from human life, but the processes that ventilate it are tied to the same boundary regions that support coastal ecosystems and fisheries.

A final concrete image helps keep the energy story honest. Imagine trying to mix a two-layer drink: dense syrup at the bottom, lighter soda above. If you stir gently at the top, you can create motion, but the layers largely persist. If you shake the glass vigorously, you can mix them, but you must supply energy, and the mixture will often re-stratify if left alone, because gravity prefers dense below light. The ocean is not a glass and it is not shaken by hands, but the principle survives at planetary scale. Winds and tides are the shaking. Internal waves are the sloshing that carries energy into the interior. Turbulence is the moment when the layers truly exchange material. Without that sequence, dense water would sink and stay sunk, and the deep ocean would become a stagnant reservoir rather than an active participant in Earth's climate.

The abyss is therefore not merely deep; it is energetically expensive. Every slow upward drift of deep water represents countless small acts of turbulence, most of them happening out of sight over rough topography, along continental margins, and in narrow passages where tides and currents are forced into acrobatics. The ocean pays for its circulation in a currency of mechanical energy, spent in microscopic whorls that, taken together, allow the heaviest waters on Earth to lighten, rise, and rejoin the surface world.

Chapter 3

Gateways to the Abyss

While the ocean covers most of our planet, the critical exchange between surface and abyss occurs in only a few specific, hostile postcodes. This chapter transports the reader to the freezing extremities of the Earth, revealing how the physics of ice formation and atmospheric cooling conspire to power the global thermohaline circulation.

3.1 The Alchemy of Ice

Seawater is an unpromising substance from which to build a solid. It is restless, chemically busy, crowded with dissolved ions, and perpetually nudged by waves. Yet in polar winter it performs a kind of everyday miracle: it turns into a floating lid. That lid is not just a passive cap. It is an active chemical filter, a factory that sorts water from salt and, in doing so, manufactures a heavy fluid capable of sinking thousands of metres. The deep ocean is filled, in part, by the by-products of freezing.

The trick begins at the scale of molecules. Liquid water is a tangle of fleeting hydrogen bonds. When it freezes, those bonds reorganize into a more regular arrangement: an ice lattice. The geometry of that lattice is fussy. It likes water molecules and not much else. Sodium and chloride ions (the familiar Na^+ and Cl^- of table salt) do not fit comfortably into the crystalline architecture. As the solid phase grows, the ice pushes the salt aside.

If freezing happened infinitely slowly, the ice might manage to segregate almost all the impurities into the remaining liquid, like a careful cook clarifying stock. In the real polar ocean, freezing often proceeds quickly: wind strips heat away, frazil crystals appear, pancakes of young ice jostle together, and sheets of new ice thicken from below. Under these conditions, salt is excluded *en masse* rather than delicately removed. Some of it is trapped temporarily in pockets and veins of salty liquid within the ice, but a large fraction is expelled downward into the water beneath.

This expulsion is called *brine rejection*. The phrase can sound mild, as if the sea were politely declining an ingredient. In fact it is a decisive push. The rejected brine is colder than the surrounding seawater and saltier too, and both changes increase density. In the polar ocean, density is destiny: the heavier fluid wants to sink. The result can be a downward pulse of buoyancy loss that drives vertical mixing and, in the right settings, initiates the formation of waters dense enough to supply the abyss.

A simple kitchen experiment captures the essence. Freeze a glass of seawater (or, more safely, salty water mixed at home). The first ice that forms is relatively fresh. The remaining liquid becomes ever saltier. Taste is not a recommended method, but conductivity meters and careful measurements confirm the separation. Early polar scientists made the same discovery with more hardship. During the late nineteenth-century drift of Fridtjof Nansen's *Fram* across the Arctic Ocean (1893–1896), the expedition collected meticulous observations of sea ice and the water beneath it. Nansen and his colleague Bjørn Helland-Hansen later helped turn these field notes into a physical picture of how the polar ocean stratifies. They could not yet watch brine plumes with modern instruments, but they grasped the central point: freezing does not merely remove heat; it rearranges salt, and that rearrangement alters circulation.

The separation is never perfect, and the imperfection matters. Newly formed sea ice contains a labyrinth of brine inclusions. At warmer temperatures (for sea ice, that can mean only a few degrees below 0°C), these inclusions are connected, like the pores in a sponge. Brine can drain through them under gravity. As the ice cools further, the brine pathways narrow and pinch off, trapping pockets of very salty fluid. This is why first-year sea ice can be surprisingly salty and why its properties evolve through the season: as temperatures fall and rise, the internal plumbing opens and closes.

Down at the ice—ocean interface, the physics is less subtle. The water immediately beneath growing ice becomes saltier as brine is rejected. Saltier water is denser, but cold water is also denser—up to a point. Near the freezing temperature, small changes can have large effects because the density difference needed to start sinking is not enormous; what matters is whether the surface layer becomes just heavy enough to overcome the stabilizing tendency of the water below.

Here is the crucial nuance: in much of the ocean, cooling alone can increase density and promote sinking, but freezing adds a second lever. Cooling changes temperature; brine rejection changes salinity. Temperature can be restored quickly in spring and summer by sunlight and warm winds, but salinity changes can persist longer, especially when they are injected downward into deeper layers. Brine rejection is therefore a way of giving winter a longer reach.

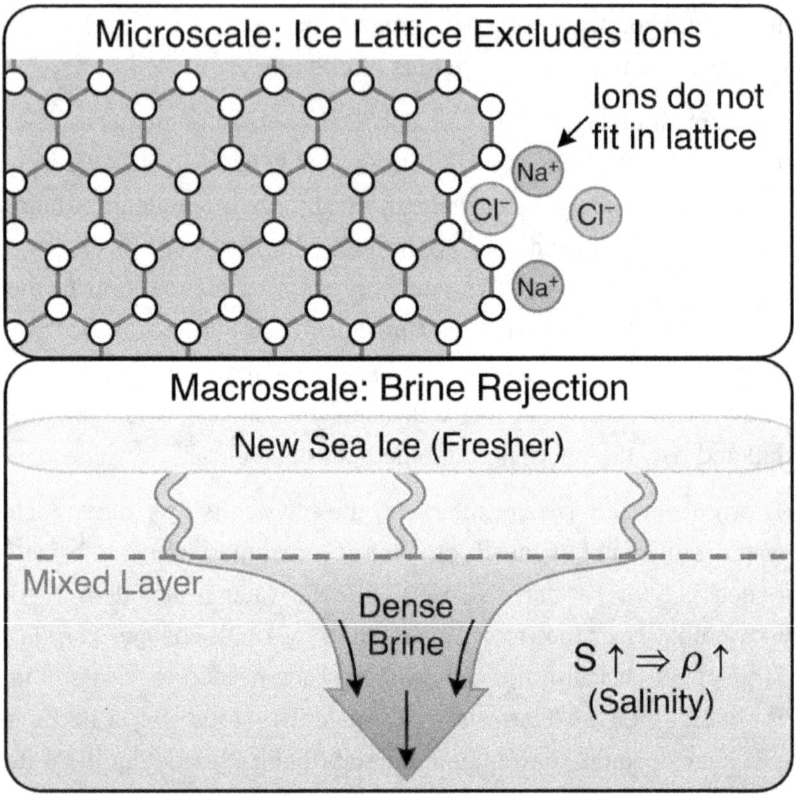

The reason this matters for the deep ocean is that brine rejection does not have to create an *extremely* salty fluid to be effective. It only has to create a fluid that is slightly denser than what sits below. Once a patch of surface water becomes heavy enough, it begins to sink, pulling more water down with it and drawing surrounding water in to replace it. That motion is convection: the same basic mechanism that stirs soup on a stove, but slowed by the vastness of the ocean and complicated by Earth's rotation.

There is also a geographical twist. In many polar regions, sea ice does not form as a quiet, uniform sheet. It forms where the ocean is exposed to bitter air and where winds keep sweeping new ice away as soon as it is made. These are polynyas: recurring areas of open water or thin ice within an otherwise ice-covered sea. To stand on the edge of a winter polynya (few people ever do, because the conditions are brutal) would be to feel the atmosphere and ocean in direct confrontation. Open water loses heat to the air far faster than thick ice does. The response is frantic freezing, and frantic freezing means intense brine rejection.

Along parts of the Antarctic coast, katabatic winds—cold, dense air draining down from the high interior of the continent—blow seaward with astonishing persistence. They push sea ice away from the shoreline and keep coastal polynyas open. Each new skim of ice that forms is exported and replaced by open water again. The ocean, in effect, is forced to keep freezing the same patch repeatedly, like a conveyor belt. The brine produced by this repeated freezing can accumulate on the continental shelf as a dense, salty layer known as dense shelf water. This is one of the essential precursors of Antarctic Bottom Water, the heaviest water mass in the global ocean, which spreads into abyssal basins and helps set the deep stratification of the planet.

The Arctic has its own versions of this process, though the outcome differs because the geography and background salinity differ. In the Arctic Ocean, large river inputs freshen the surface, making it harder for surface waters to become dense enough to sink deeply. Brine rejection still occurs, and it still drives local mixing and contributes to the formation of cold, salty waters on shelves, but the pathways into the global abyss are more restricted. In the Antarctic, by contrast, the shelves ring an open connection to the deep ocean, and the densest shelf waters can descend the continental slope, entraining surrounding waters as they go.

One might imagine the rejected brine as a narrow jet, a clean arrow of purple ink sinking into blue water. Reality is messier, and the messiness is part of the power. Brine is released patchily: from brine channels, from leads where new ice forms, from frazil crystals that grow in supercooled water and rise or fall depending on turbulence. Each release creates small-scale density anomalies. These anomalies generate fingers and plumes that mix with surrounding water as they descend. Mixing matters because it sets how much dense water is produced and what its final properties are. A plume that entrains too much lighter water will lose its ability to sink; a plume that remains relatively unmixed can punch deeper.

Even when the plumes are small, their collective effect can be large. Over weeks to months, repeated brine rejection can erode the stable layering of the upper ocean. In a strongly stratified water column, the surface is light and the deeper water is heavy, and vertical exchange is suppressed. Brine rejection is one of the few processes capable of pushing the surface in the other direction: making it heavy, destabilizing the column from above, and thereby enabling vertical exchange on scales that matter for climate. This exchange ventilates deeper layers with oxygen, exports surface properties downward, and rearranges heat and salt in ways that persist beyond the season.

There is a human-scale reason this polar chemistry deserves attention. The deep ocean is Earth's largest long-term heat reservoir accessible on human timescales, and it is also a major reservoir for carbon. Water that sinks from the surface carries with it not only its temperature and salinity but also dissolved gases and, increasingly, anthropogenic carbon absorbed from the atmosphere. The rate at which the ocean interior is renewed sets the pace at which it can store heat and carbon, and also the pace at which deep waters are replenished with oxygen. Brine rejection is one of the mechanisms that opens the gate.

Because the process is both common and difficult to observe, it has a peculiar scientific history: everyone who works in polar oceans knows it must be happening, yet it is hard to watch directly. Winter darkness, moving ice, and lethal winds make routine measurements rare. Much of what is known comes from clever combinations of field campaigns, laboratory studies of sea ice physics, moorings that can survive under ice, and satellite observations that infer ice production from heat loss and ice motion. Over time, these lines of evidence have converged on a coherent picture: the ocean does not simply cool into the abyss; it is *salted* into it.

That coherence does not make the system simple. A small change in the balance of freezing and melting can flip the sign of the buoyancy forcing. If, for example, a region experiences more melting of glacial ice or more precipitation, the freshening can strengthen stratification, making it harder for brine rejection and cooling to trigger sinking. Conversely, stronger winds that open more leads can increase ice production, increasing brine rejection and densification. The same surface patch can behave like a pump in one decade and like a lid in another.

There is also an irony at the heart of the story: sea ice floats, yet its creation can help fill the ocean's deepest layers. The floating lid is an engine precisely because it does not incorporate what makes seawater heavy. The salt left behind is the weight. It pools, it spills, it sinks, and it carries the fingerprint of polar winter downward into darkness, where it can persist long after the surface has returned to summer light.

3.2 Chimneys of the North

If sea ice is a chemical sieve, the North Atlantic in winter is a test of nerve. Here the atmosphere does not merely cool the ocean; it tries to wrestle heat from it as fast as physics allows. Wind tears at the surface, spray freezes on railings, and the sea answers with steam-like plumes of evaporation. In a few favoured places the surface waters become dense enough to sink rapidly, drawing lighter water down behind them in narrow, intense columns. Oceanographers call these episodes *open-ocean deep convection.* The columns are often described as chimneys, and the metaphor is apt: they are transient vertical pipes through which the surface communicates directly with the deep.

The material produced is North Atlantic Deep Water (NADW), a family of cold, relatively salty waters that occupy much of the Atlantic below about a kilometre depth. NADW is not a single uniform substance; it is a blend, stitched together from different source regions and modified along the way. What unites those sources is the same requirement that governs all deep waters: the surface must be made heavy enough to sink. In the North Atlantic, this heaviness is created mainly by cooling and evaporation, with an important supporting role played by salt imported from lower latitudes and, in certain settings, by brine rejection from sea-ice formation.

A useful way to picture the problem is to think like the ocean. The surface layer is usually lighter than the water beneath it, which keeps the water column stable and discourages vertical mixing. To overturn that stability, the ocean must lose buoyancy at the surface. Cooling removes buoyancy because cold water is denser than warm water. Evaporation removes buoyancy because it leaves salt behind, raising salinity and density. Both processes are strongest in winter, when cold, dry air masses sweep off continents and over the sea.

Yet winter alone is not enough. Deep convection is famously intermittent, and some winters that feel severe on land produce surprisingly modest sinking at sea. The difference lies in *preconditioning*: the state of the ocean at the start of winter. If the upper ocean is already weakly stratified, a given amount of cooling can mix it deeply. If the ocean is capped by a fresh surface lens, the same cooling may only stir the top few hundred metres. The North Atlantic is therefore a place where memory matters. What happened last year, and the year before, can set the stage for what winter can accomplish now.

Two regions have been central to this story: the Nordic Seas (the Greenland, Iceland, and Norwegian seas) and the Labrador Sea. They sit like cold basins and marginal seas around the northern rim of the Atlantic, positioned at a crucial junction between the Arctic and the temperate ocean. They are also places where geography forces decisions. Ridges and sills in the seafloor act as thresholds. Narrow passages funnel currents. Some waters are trapped and transformed locally, while others spill over submarine divides and slide into the deep Atlantic, carrying their properties southward.

Start with the Nordic Seas, because they make the North Atlantic feel truly polar. Here warm, salty Atlantic water flows northward, cools, and meets the colder, fresher influence of the Arctic. Sea ice forms seasonally along the margins, especially toward Greenland and in the Arctic gateways. As the Atlantic water loses heat to the atmosphere, it grows denser, and in places it can sink to intermediate and deep depths. The deepest contributions to NADW from this region are often associated with overflows: dense waters that collect in the Nordic basins and then pour southward across the Greenland–Scotland Ridge through a few key gaps. After passing these sills, the overflows descend the continental slope in turbulent plumes, entraining surrounding water and thickening into broader currents. The drama occurs out of sight, but its imprint is detectable far downstream in the temperature and salinity of the deep Atlantic.

The Labrador Sea, by contrast, is less constrained by sills and more exposed to the open subpolar Atlantic. It sits between Greenland and northeastern Canada, a bowl surrounded by steep continental slopes. In winter it can be assaulted by cold, dry winds streaming off the ice-covered land and sea ice. Those winds strip heat from the surface and can drive the mixed layer to astonishing depths. When this happens, the water column that was previously layered becomes nearly uniform over a kilometre or more: temperature and salinity are mixed vertically, and the ocean interior is ventilated. The product is Labrador Sea Water, an important component of the broader NADW mixture. Not all deep convection in the region occurs in the Labrador Sea itself, but the Labrador has become a symbol of how abruptly the ocean can connect surface to depth.

The phrase *chimney* is not just poetic; it describes an identifiable dynamical structure. Convection begins with surface buoyancy loss. Cooling and evaporation increase density in the surface mixed layer until it matches and then exceeds the density below. At that point, small instabilities grow: denser parcels sink, lighter parcels rise, and the mixed layer deepens. If the forcing continues, the deepening can accelerate, and localized plumes can punch downward.

Earth's rotation complicates this picture in a way that is easy to miss from a shore-based intuition. A non-rotating fluid can overturn freely, but in the ocean the Coriolis effect tends to organize motion into swirling, horizontally balanced flows. Deep convection therefore cannot be imagined as a simple vertical waterfall. Instead, narrow plumes form within a larger, rotating mixed patch. After the strongest winter forcing ends, the mixed patch does not simply sit there. It restratifies: eddies peel off from its edges, stirring in lighter water from the surroundings and tilting density surfaces back into place. The newly formed dense water is then exported laterally, slipping away from the formation site into boundary currents and the broader subpolar gyre. Much of the ocean's work is done in this aftermath: a short, violent period of mixing followed by months of adjustment and spreading.

A concrete historical moment shows how sensitive these chimneys can be to freshening. In the late 1960s a remarkable event unfolded in the North Atlantic: the Great Salinity Anomaly. A pulse of unusually fresh water, traced in part to Arctic outflows, propagated around the subpolar gyre. In the Labrador Sea, this freshening helped cap the surface. For several winters, from 1969 through 1971, deep convection there largely shut down. The atmosphere could still cool the surface, but the fresh layer acted like a buoyant lid; the water column resisted being overturned. The episode is remembered not because it was the only time convection weakened, but because it showed how quickly a key site could flip from an active chimney to a quiescent stratified basin. Recovery came abruptly in 1972, when strong atmospheric forcing and changes in the inflow of saltier waters combined to erode the cap. The message is uncomfortably clear: the system has thresholds, and crossing them can change the deep ocean's supply line on timescales of years, not centuries.

That story resonates now because the North Atlantic is one of the places where climate change can act on both levers at once. Warming reduces winter heat loss and can make the surface harder to densify by cooling. At the same time, increased freshwater input—from precipitation, river runoff, and melting of glaciers and ice sheets—can lower salinity, strengthening stratification and further suppressing sinking. The result is not necessarily a smooth decline. Because convection depends on both forcing and preconditioning, it can lurch: a period of shallow mixing, a sudden deep winter, then a return to suppression, all while the longer-term background shifts.

Modern observations have made the story richer, and in some ways less tidy. For decades a simple narrative was popular: the Labrador Sea is the main throttle controlling the deep limb of the Atlantic overturning circulation. That narrative had the virtue of clarity and the support of dramatic wintertime deep mixing events. But the ocean rarely honours single-cause explanations. A major observational program, OSNAP (the Overturning in the Subpolar North Atlantic Program), began sustained measurements across the subpolar Atlantic in the mid-2010s. Analyses published by M. Susan Lozier and colleagues in 2019 emphasized that variability in subpolar overturning is strongly associated with water-mass transformation and export in the Irminger and Iceland basins, not solely with Labrador Sea deep convection. In other words, the Labrador Sea matters, but it is not the only gateway, and not always the dominant one.

This matters for how to interpret the word *chimney*. It is tempting to picture a single, reliable vertical pipe. In reality, there are multiple sites where winter forcing can trigger deep mixing, and there are multiple routes by which dense waters leave those sites. Convection near the southern tip of Greenland, southeast of Cape Farewell, has produced repeated deep mixing events in the Argo float era, illustrating that the most active chimneys can migrate with changes in currents and stratification. The ocean's gateways are not fixed doors in a wall; they are more like recurring weaknesses in a shifting field of ice, wind, and current.

A second modern nuance concerns timing. People often assume that the coldest winter produces the deepest convection. Observations suggest it is not so simple. In the Labrador Sea during the 2010s, for example, exceptionally strong cooling occurred in 2015, yet the deepest convection in that decade was observed later, around 2018, consistent with the idea that preconditioning across multiple winters can deepen the mixed layer even if the most intense surface cooling happens earlier. The ocean can spend a few winters gradually eroding its stratification, and then one winter finally breaks through. That is not merely an academic curiosity; it is the kind of delayed response that can surprise societies used to linear cause and effect.

To keep the physics grounded, it helps to anchor the discussion in a specific scene. Imagine standing on the deck of a research ship in the Labrador Sea in midwinter. The horizon is low, the light short. The wind is a physical force, and the ocean surface is a patchwork of whitecaps and spindrift. Instruments lowered over the side record the changing density structure: the mixed layer deepening over days and weeks, temperature and salinity profiles flattening. If the forcing is strong enough and the stratification weak enough, the mixed layer can reach depths comparable to a mountain's height. When the storm passes, the surface may briefly look calmer, but the ocean's interior has been rearranged. Oxygen has been injected downward; carbon absorbed at the surface has been carried with it. A surface signature has become a deep one.

The Nordic Seas offer a different kind of scene: less a single basin-wide mixing event and more a tapestry of transformations. Warm Atlantic water enters, releases heat to the atmosphere (helping keep parts of northern Europe milder than their latitude would suggest), becomes denser, and is modified by interactions with sea ice and freshwater outflows. Dense waters formed there can accumulate and then spill southward through the Denmark Strait and the Faroe Bank Channel, among other passages. These overflows are some of the most energetic deep currents on Earth, and their descent is intensely turbulent. That turbulence is not a detail. It determines how much surrounding water is entrained and therefore how the overflow's temperature and salinity are diluted before it joins the deep Atlantic. In climate terms, entrainment sets the properties of the deep water that spreads through the basin and influences how heat and carbon are stored at depth.

Both regions share a vulnerability: a little extra freshwater at the surface can prevent sinking by lowering density and strengthening stratification. The sources of that freshwater differ. In the Labrador and Irminger seas, freshwater can arrive via sea-ice export and melt, precipitation, and runoff from Greenland. In the Nordic Seas, Arctic outflows and sea-ice processes play a stronger role. But the mechanism is the same: fresh water is buoyant. It is a kind of anti-weight placed at the surface.

There is a subtle feedback hidden in this. If convection weakens, less warm water may be brought into contact with the cold atmosphere, potentially reducing local heat loss. At the same time, changes in surface temperature can alter how effectively cooling translates into density change because the thermal expansion coefficient of seawater varies with temperature and salinity. Near the freezing point, the density response to temperature is different than it is in warmer waters. This means the ocean's sensitivity to winter forcing can change as the system shifts, making it harder to reason by intuition alone.

The human stakes of these northern chimneys are often described in terms of the Atlantic Meridional Overturning Circulation (AMOC), the large-scale pattern in which warm surface waters flow northward, release heat, and return southward at depth. NADW formation is a key component of that return flow. Changes in the rate or properties of deep water formation can therefore influence regional climate, sea level along coastlines, marine ecosystems, and the ocean's role as a sink for heat and carbon. But the most immediate lesson from the chimneys is not a single prediction; it is a way of thinking. The deep ocean is ventilated through a small number of intense events in a small number of places, and those events depend on a delicate balance between atmospheric violence and the ocean's layered structure.

When winter succeeds in breaking the stratification, the result is startlingly direct: the surface communicates with the deep in a matter of weeks. When it fails, the deep ocean is starved of renewal, and the surface retains its buoyant cap. The difference between those outcomes can hinge on a few tenths of a unit of salinity or a degree of temperature, changes small enough to seem trivial until they decide whether a water parcel will stay afloat or begin its long descent into the Atlantic interior.

3.3 The Southern Heavyweight

The deepest ocean is not filled from the top down in a uniform drizzle. It is filled in pulses and pours from a handful of places where winter is allowed to be extreme. Antarctica is the most consequential of those places, not because it is colder than anywhere else on Earth in some absolute sense, but because its geometry and winds collaborate to manufacture a water mass heavier than any other: Antarctic Bottom Water (AABW). Once formed, it creeps and floods into the abyssal basins of every ocean except the Arctic, setting the deep ocean's baseline temperature, salinity, and oxygen supply.

AABW begins its life on the continental shelves ringing Antarctica, in waters shallow enough to be stirred by tides and to feel the direct grip of the atmosphere through openings in the sea ice. The paradox of AABW is that it is made in relatively shallow places yet ends up occupying the deepest. The mechanism that bridges that gap is a sequence of density-making steps, each one conditional and fragile: create dense shelf water, allow it to spill off the shelf, let it entrain and thicken without becoming too diluted, and finally feed it into the deep boundary currents that spread it around the planet.

The ingredients are familiar by now. Cooling increases density. Brine rejection during sea-ice formation increases density more persistently by raising salinity in the underlying water. In the Antarctic, those two processes are sharpened by a distinctive production line: coastal polynyas maintained by katabatic winds.

A polynya is an area of open water or thin ice within the sea-ice cover. Along many Antarctic coasts it is not a curiosity but a recurring feature, reappearing each winter like a stage set rebuilt in the same place. The reason is the wind. Katabatic winds are gravity-driven outflows of cold, dense air that form over the high Antarctic interior and accelerate downslope toward the coast. They can be ferocious, not primarily because they are the strongest winds on Earth at any given moment (hurricanes have their own claim), but because they can be frequent, directional, and persistent. When katabatic winds blow offshore, they push newly formed sea ice away from the coast. The coastal water surface is repeatedly exposed, and each exposure is an opportunity for rapid heat loss and new ice formation.

This is why many Antarctic coastal polynyas are sometimes called "ice factories." The factory metaphor is not sentimental. Thin ice and open water lose heat far faster than thick ice. Thick ice insulates the ocean from the atmosphere; thin ice does not. An offshore wind that continually removes thin ice prevents the system from ever reaching the insulated, throttled state. The ocean is forced to keep freezing. Each round of freezing rejects brine into the surface layer, making the water beneath the polynya saltier and denser. Over the course of a winter, this can create a pool of dense shelf water (DSW) sitting on the continental shelf like a cold, salty lens pressed against the seafloor.

DSW is not yet AABW, but it is the essential seed. To become bottom water, it must leave the shelf. That departure is not guaranteed; shelves are broad, sills can block pathways, and mixing can erode density. When DSW does escape, it typically does so as a gravity current: a dense flow that spills over the shelf break and cascades down the continental slope. On its way down, it entrains surrounding waters, especially Circumpolar Deep Water that sits offshore at intermediate depths. Entrainment is both a necessity and a risk. It increases the volume of the descending plume, helping to generate a large abyssal water mass, but it can also warm and freshen the plume, potentially reducing its ability to sink to the very bottom. The AABW that finally spreads into the deep basins is thus a blended water mass, its properties shaped by both the shelf "ice factory" and the turbulent descent.

The role of the wind, in this sequence, is almost theatrical: it clears the stage again and again so the ocean can perform the same act repeatedly. Yet the wind does not work alone. Tides help mix shelf waters and can deliver heat and salt to the polynya region, affecting how much ice forms. Sea-ice dynamics matter because they determine whether ice is exported cleanly or piles up into thick ridges that reduce heat loss. The local bathymetry matters because it controls where dense water can gather and where it can spill. AABW is therefore not a single, simple product but the outcome of an entire coastal system.

A concrete example helps fix ideas. Off East Antarctica, the region near the Mertz Glacier Tongue was long recognized as a prolific site of sea-ice production and dense water formation. In 2010, a dramatic change occurred: the Mertz Glacier Tongue calved, altering the local icescape and, crucially, the way sea ice forms and is exported. When the geometry of the coastline and grounded ice changes, the wind-driven ice conveyor changes with it. Satellite-based studies in the following years documented substantial shifts in sea-ice production in the region. This is the kind of event that makes the Antarctic feel less like a static boundary and more like a living machine whose gears can slip. A single calving event can reorganize an "ice factory," changing how much brine rejection occurs and therefore how much dense shelf water is produced.

Another site, Cape Darnley, has been identified as a major AABW source region. The significance of such sites is not only local. Because AABW spreads widely, a change in one region can, over time, be detected far away in the temperature, salinity, and oxygen content of abyssal basins. Oceanographers have observed multidecadal changes consistent with a weakening of abyssal ventilation in parts of the Southern Ocean and the downstream basins, including signs of deep warming and deoxygenation. Those observations matter because the abyss is slow, and slow changes in the abyss often indicate persistent shifts in formation or spreading rather than short-lived weather noise.

The historical relationship between people and Antarctic bottom water is mostly indirect. There are no coastal cities depending on AABW for fisheries in a simple sense, and no shipping lanes that run through its formation sites in winter without extraordinary risk. But the indirect relationship is profound: AABW helps set the storage capacity of the ocean for heat and carbon, and it contributes to the oxygenation of the deep sea. When the abyss is ventilated, it receives oxygen from the surface; when ventilation weakens, oxygen declines. In the deep ocean, oxygen is not a luxury but a boundary condition for life. Many deep-sea organisms live near the edge of what oxygen levels allow. A slow reduction in abyssal oxygen can reshape habitats across enormous areas, even if the surface ocean seems unchanged to the casual observer.

One of the most intriguing, and unsettling, aspects of Antarctic bottom water formation is that it has both a robust pathway and a wildcard. The robust pathway is the coastal polynya route: wind-driven ice production, brine rejection, dense shelf water, downslope cascading, entrainment, and abyssal spreading. The wildcard is open-ocean polynyas in the Weddell Sea, particularly near Maud Rise. These are large holes that can open within the sea-ice cover far from the coast, exposing the ocean directly to the winter atmosphere. Their occurrence can enable deep convection in the open ocean, a more direct analogue to the northern chimneys, but occurring in the Southern Ocean where the background stratification and circulation differ.

In the mid-1970s, satellites revealed a remarkable event: a giant polynya in the Weddell Sea persisted for multiple winters, an opening vast enough to be seen from space at the dawn of satellite meteorology. Then, for decades, such events were absent or far smaller. In

2016 and 2017, large open-ocean polynyas reappeared in the same general region, catching the attention of scientists and the public alike. The details of how and why these openings form remain an active area of research, involving a complicated interplay between ocean heat stored at depth, sea-ice dynamics, atmospheric storms, and the local seafloor topography that can anchor rotating currents. What is clear is that when an open-ocean polynya opens, the system's rules change. The insulating lid of sea ice is removed over a broad area, and heat can be vented rapidly to the atmosphere. That can deepen the mixed layer dramatically and potentially create dense waters through convection, though whether this contributes significantly to modern AABW production relative to coastal sources is still debated.

The existence of this wildcard matters for how one thinks about stability. A coastal polynya is maintained by winds that can be measured and modelled, and its ice production can be estimated from satellite observations with increasing skill. An open-ocean polynya is more episodic and less predictable, and because it can reorganize air-sea exchange over a wide area, it can have outsized impacts when it occurs. The Southern Ocean is a region where rare events can matter.

AABW is sometimes described as the "lower limb" of global overturning circulation, while NADW is a "deep limb" of the Atlantic overturning. Those phrases can sound like plumbing, but the physics is more like a balance of weights and pathways. AABW is heavier than NADW, and that ordering matters: in many basins, AABW occupies the deepest layers while NADW sits above. If AABW formation weakens, the deepest layers can warm and freshen, and the interface between water masses can shift. Because the abyssal ocean is vast, even small changes in its properties imply large amounts of heat and freshwater redistributed.

There is also a feedback that is easy to state and hard to live with. The production of dense shelf water relies on salt being concentrated by sea-ice formation. But many Antarctic shelf regions are being freshened by increased meltwater input and changes in precipitation and sea-ice processes. Freshening makes it harder to form very dense waters because it lowers the starting salinity and strengthens the stratification that resists mixing. In the simplest picture, add freshwater and you reduce the density of shelf waters; reduce density and you reduce the propensity for those waters to sink into the abyss. Assessment-level syntheses, including those of the IPCC, have judged there to be medium confidence that AABW formation and export will decrease under continued warming and freshening. "Medium confidence" is not a shrug; it reflects the combination of strong physical reasoning, growing observational evidence, and persistent uncertainty about the relative roles of different regions and processes.

The uncertainty is not trivial because the Antarctic margin is not one uniform factory line. Some shelves are broad, some narrow. Some are sheltered by ice shelves, others exposed. Some polynyas produce enormous amounts of ice; others are smaller but strategically placed near pathways off the shelf. The resulting bottom water is also not uniform; different source regions create bottom waters with slightly different temperatures, salinities, and chemical signatures, which then mix as they spread.

A final point brings the science back to something almost tactile. Dense water formation is often described with abstractions: fluxes, stratification, buoyancy. But on the Antarctic shelf, density has the feeling of weight. Salt rejected into the ocean is not just a number; it is a heaviness that collects near the bottom and waits for a chance to spill. When the spill happens, it is not a gentle diffusion but a current with momentum, hugging the seafloor, thickening as it swallows surrounding water, and then turning into a slow, persistent flow that can be traced across an ocean basin. The abyss is filled by these patient, bottom-hugging movements, each one beginning with winter air and wind forcing a patch of ocean to keep freezing, over and over, until the water becomes heavy enough to fall.

Chapter 4

The Battle for Buoyancy

The deep ocean's heartbeat depends on a singular, fragile property: the density of surface water in the North Atlantic. While cooling drives water down, an influx of fresh water-from intensifying rains and melting ice-acts as a buoyant shield, threatening to seal off the ocean's interior. We investigate how this "freshwater forcing" destabilizes the delicate balance of salt and temperature that keeps the global circulation in motion.

4.1 The Great Freshening

The ocean's surface looks uniform from an airplane window: a moving sheet of steel-blue, lightly wrinkled by wind. Up close, it is anything but uniform. Seawater has a taste, and that taste varies. The word *salinity* simply means how much dissolved salt the water contains, usually expressed as grams of salt per kilogram of seawater (or, in practice, as a closely related "practical salinity" number). The global average sits near 35, but the number is less important than the contrast: salty regions and fresh regions, and the boundaries where the two meet.

Those contrasts matter because salinity, alongside temperature, helps set seawater density. A small change in salinity can be a large change in buoyancy when the water is near freezing. In the high latitudes, where winter air is cold enough to pull heat from the sea at a furious pace, salinity often plays the deciding role: whether surface water becomes heavy enough to sink, or light enough to float stubbornly at the top.

Freshening is the straightforward part: the surface ocean becomes less salty. The less straightforward part is how a warming world can make that happen in several ways at once, some local and some remote, some gradual and some arriving like a sudden delivery. The phrase “the hydrological cycle is intensifying” sounds abstract until it is translated into everyday experiences: heavier downpours, more moisture in the air, rivers running higher, snowpacks behaving differently, and ice once locked on land turning into runoff. All of that extra freshwater ends up somewhere. Much of it ends up in the sea.

A useful rule of thumb is that warm air is a better sponge than cold air. As the atmosphere warms, it can hold more water vapor. That does not mean it rains everywhere all the time. It means that the atmosphere can pick up more moisture where evaporation is strong and release more moisture where storms and rising air wring it out. The result is a sharpening of the planet’s wet-and-dry pattern: “wet gets wetter, dry gets drier” is an oversimplification, but it captures the broad tendency. High latitudes, particularly around the Arctic, are among the places where precipitation and runoff have increased. There is more freshwater input from above, delivered by storms and snowfall, and more from the sides, delivered by rivers draining vast northern watersheds.

The Arctic Ocean is especially good at storing this freshwater, because it is built like a basin with a lid. A strong halocline—layers of relatively fresh water over saltier Atlantic-derived water—helps keep the surface buoyant and stable. Winds can then corral the fresh surface layer into large reservoirs. The most famous is the Beaufort Gyre, a slow, wind-driven whirlpool north of Alaska and Canada.

Picture it as a savings account for freshwater: when winds are arranged in a certain persistent pattern, they push the surface water toward the center of the gyre, piling it up slightly and trapping freshwater there. The "pile" is subtle—measured in tens of centimeters of sea surface height—but spread over an oceanic area it amounts to an enormous volume.

This matters far beyond the Arctic map because the freshwater does not stay put forever. It leaks. It leaks through the Canadian Arctic Archipelago into Baffin Bay and the Labrador Sea; it leaks through Fram Strait east of Greenland into the Nordic Seas. The timing is not regular like a tide. It depends on winds, ice conditions, and the state of ocean currents. In other words, the Arctic can store freshwater for years and then export a pulse when the system's gates swing open. High-latitude North Atlantic regions-precisely the places where winter cooling can drive surface water to sink-feel these pulses as sudden injections of buoyancy.

One of the best historical illustrations is the episode that oceanographers still discuss with a kind of wary respect: the Great Salinity Anomaly of the late 1960s through early 1980s. Ships crossing the northern North Atlantic began finding surface and upper-ocean waters that were fresher than expected across a wide region. The anomaly did not simply appear and vanish locally; it traveled, circulating around the subpolar gyre like dye carried by a slowly turning wheel. It showed up in one area, then later in another, and it lingered long enough to affect winters of deep mixing. It was, in essence, a lesson in how the ocean can carry a memory of freshwater for more than a decade.

The details are complex-no single anecdote captures the whole event-but one concrete image is simple: for a time, the upper ocean in parts of the subpolar North Atlantic was missing salt. That missing salt is missing density. When winter storms tried to stir and cool the surface layer, they had to work against an extra measure of buoyancy. Even where convection still occurred, the conditions that *prepare* a region for deep mixing were altered: how thick the mixed layer became in autumn, how strongly stratified the water column was at the start of winter, how readily surface cooling could reach deeper layers. A decadal-scale freshening is not just a change in taste; it is a change in how winter behaves in the ocean.

Freshening also arrives by a more direct and emotionally charged route: melting land ice. Greenland's ice sheet is not sea ice. When sea ice melts, it is already floating and its meltwater does not raise sea level. Greenland's ice is on land; when it melts and runs into the sea, it adds water and changes both sea level and salinity. The freshwater is delivered seasonally-summer meltwater pulses are the most visible-but the long-term trend is what matters for buoyancy: year after year, a greater input of low-salinity water enters the surrounding seas.

It is tempting to imagine that this freshwater instantly dilutes the entire North Atlantic, like pouring a glass of water into a pot. The real ocean does not oblige. Freshwater often stays near the surface and near the coast at first, guided by boundary currents and shaped by winds. It can form thin lenses and filaments that are hard to track. Yet those near-surface layers are exactly where winter atmosphere—ocean exchange operates: heat leaves the sea through the surface, and the surface is where winds do their stirring. A small salinity change concentrated in the top tens of meters can matter more for buoyancy than a larger change smeared through a kilometer of water.

Rivers provide a third route, and they can act like a steady faucet rather than a seasonal pulse. The great northern rivers—Ob, Yenisei, Lena, Mackenzie—deliver a tremendous volume of freshwater to the Arctic Ocean. Their discharge is shaped by precipitation, snowmelt timing, and the thawing of soils and permafrost. In a warmer climate, river runoff patterns can shift in ways that reinforce Arctic freshwater storage: earlier spring melts, altered storm tracks, and changing ground conditions can all change how much freshwater reaches the sea and when. Again, the Arctic is not a simple pass-through. It stores, mixes, freezes, and releases.

Against this freshwater push is the ocean's salt supply line. The North Atlantic receives relatively salty water from the subtropics, carried northward by currents that are part of the broader overturning system. Salty inflow is not a decorative detail; it is part of what primes high-latitude waters to become dense enough to sink once cooled. When the balance shifts—more freshwater added at the top, or less salt delivered from the south—the upper ocean becomes harder to densify.

Two misunderstandings are common here. The first is the belief that "warming" and "freshening" are the same story. They often travel together, but they affect buoyancy differently. Warming reduces density, freshening reduces density, but not always with equal leverage. Near freezing, temperature changes have less room to move density (because water is already cold), while salinity can still change density appreciably. That is why a modest freshening can sometimes have an outsized impact on winter convection in near-freezing waters.

The second misunderstanding is that freshwater is always a gentle, continuous trend. In reality, freshwater forcing can arrive as variability superimposed on a trend: a background intensification of the hydrological cycle, punctuated by episodic releases from Arctic storage, extreme years of sea-ice melt, unusual wind patterns that reroute currents, or anomalous melt seasons on Greenland. The ocean does not respond like a thermostat turning smoothly; it can respond like traffic: mostly flowing, then suddenly backing up when conditions align.

A vivid modern example is the unusual freshening observed in the eastern subpolar North Atlantic in the 2010s. Oceanographers tracking long records of salinity found a pronounced drop over a few years-remarkable not only for its magnitude but also for its geography, appearing strongly in the eastern basins that are normally under the influence of the warm, salty North Atlantic Current. The episode drew attention to a subtle point: you do not need to add freshwater directly above a region to freshen it. You can freshen it by changing the pathways-by diverting Arctic-origin freshwater into a different route, or by weakening the import of salty subtropical water that normally keeps the region dense. Winds matter because winds steer the surface ocean, and the surface ocean decides where freshwater ends up.

That steering is part of the human-scale story. The North Atlantic is bordered by societies that depend on predictable seasons and stable coastlines. When the subpolar gyre freshens and warms or cools in unusual ways, fisheries can shift and ecosystems can reorganize. A few tenths of salinity do not sound like much, but they are tied to changes in stratification, nutrient supply, and the timing and depth of mixing-factors that ripple into the food web. For coastal communities, the ocean's physical changes are rarely the headline, but they can be the quiet mechanism behind the headline.

The freshwater itself also changes sea level locally. Freshwater is lighter than salty water. If you add a cap of fresher water at the surface, the sea surface can rise slightly in that region simply because the same mass occupies a slightly larger volume and sits higher due to lower density. This is not the same as global mean sea level rise from added water, but it matters for regional sea level patterns. In the Labrador Sea, for example, persistent changes in density structure can leave a signature in sea surface height that satellites can detect. To someone living on a coast, a few extra centimeters superimposed on storms and tides is not academic.

Freshening is therefore both a climate signal and a dynamical actor. It tells us that the water cycle is changing, and it also participates in changing the ocean circulation that helps set climate. The most consequential place where those roles overlap is where the North Atlantic makes deep waters-regions where winter cooling and brine rejection from sea-ice formation can densify surface waters enough for them to sink and ventilate the deep ocean. Freshening does not need to stop convection everywhere to matter. It can make convection shallower, rarer, or more localized; it can shift the "sweet spot" of where dense water forms; it can change the properties of the water that sinks, which then changes the deep ocean's temperature and salinity downstream.

One way to keep the story grounded is to imagine two glasses. In one glass is cold, salty seawater; in the other is cold, slightly fresher seawater. Both are cold enough that winter air can keep pulling heat out. But only one has the extra density from salt that allows it to sink readily when chilled and stirred. The difference between the two glasses is invisible at a glance, yet it can determine whether the ocean's vertical exchange opens like a trapdoor or stays shut.

The Great Freshening is not a single event. It is an emerging pattern: a high-latitude ocean increasingly influenced by freshwater inputs from rain, rivers, and ice, and increasingly shaped by the atmosphere's ability to transport and deliver that freshwater in bursts. The ocean has always lived with variability-storms, oscillations, shifting winds. What is new is the background state on which that variability rides: an atmosphere with more moisture to spend, an Arctic with altered sea ice, and an ice sheet adding freshwater to the ocean at a growing pace. Buoyancy is the currency that decides whether surface waters can join the deep ocean. Freshening spends that currency, one storm and one melt season at a time.

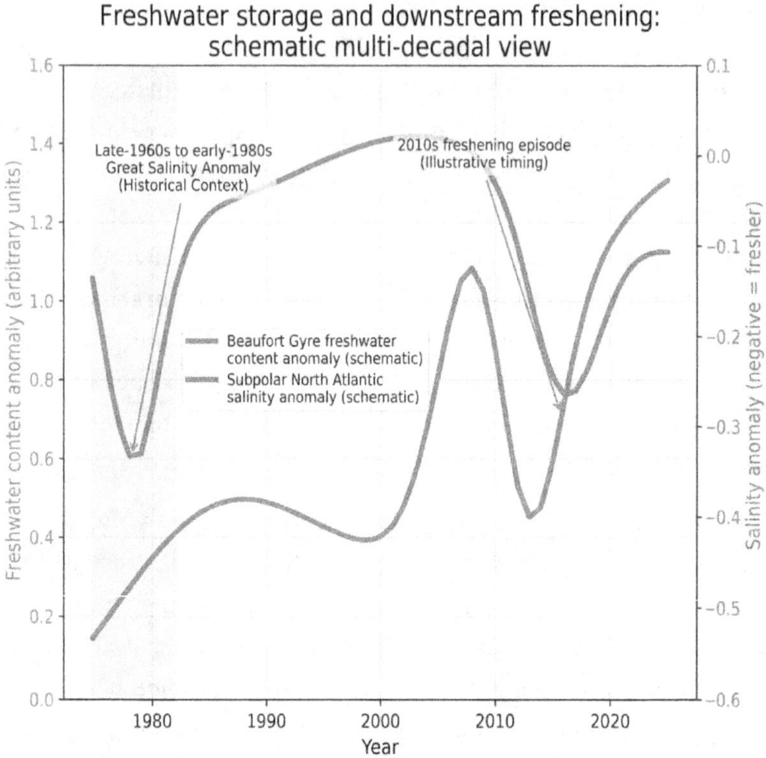

4.2 A Lid on the Ocean

Stand on the coast of Greenland in late summer and the meltwater makes itself heard before it is seen: a busy sound of trickles joining into streams, streams dropping into gullies, gullies becoming rivers that were not there a generation ago. In places, the water disappears into the ice itself, vanishing down vertical shafts called *moulins* that drill through the glacier like drainpipes. However it travels, the ending is similar. Freshwater reaches the sea, and when it does, it behaves less like a dye and more like a material with a stubborn preference for the surface.

The preference is physical. Freshwater is lighter than seawater. If you pour fresh water into a tank of salt water, it spreads across the top, forming a layer. The interface can be razor sharp in a laboratory and more ragged in the real ocean, where waves and wind stir it, but the tendency remains: the fresher water wants to float. Oceanographers call the resulting layering *stratification.* A stratified water column is one in which density increases with depth, so that lighter water sits above heavier water. That sounds stable because it is stable: the water resists being mixed vertically, just as oil and vinegar resist being stirred into one another once the spoon stops.

When stratification is strong enough, it acts like a lid. The phrase is not poetic license; it describes a genuine barrier to vertical exchange. The ocean moves heat upward mainly by mixing. If surface water can be stirred down and deeper water can be brought up, heat stored at depth can vent to the atmosphere in winter storms. If the surface is capped by a buoyant, fresh layer, the stirring is blunted. Wind may ruffle the surface, but the energy is spent sloshing around in the top tens of meters rather than punching downward. Winter air may be cold enough to freeze skin on puddles, but it cannot easily extract heat from below the cap. The deeper ocean becomes insulated by buoyancy.

This is the part of the story where it helps to be precise about what controls whether surface water sinks. Cooling makes water denser, and in much of the ocean that is enough: winter chill creates heavy surface water that slips downward. Near the freezing point, however, the density "reward" for cooling becomes smaller. Seawater cannot cool far below 0°C without freezing, and as it approaches its freezing point the density change per degree of cooling diminishes. Salinity then becomes the lever with more range. A small reduction in salinity can cancel the densifying effect of a large burst of cooling. That is why a thin freshwater layer can dictate the entire winter outcome:

sink or float, mix deep or mix shallow.

Greenland meltwater is a particularly efficient lid-maker because it arrives where the ocean is already near a tipping balance between mixing and stratification. Much of the melt drains into the seas around Greenland that feed the deep-convection regions of the Labrador Sea and the Nordic Seas. These are among the North Atlantic's critical "ventilation windows," where winter storms and heat loss can produce very deep mixed layers and, in some years, dense waters that become part of the lower limb of the Atlantic overturning circulation. In such places, the difference between a water column that is barely mixable and one that is stubbornly stratified can be a difference of only a few tenths of salinity at the surface.

The meltwater does not need to spread everywhere to matter. It matters most where it is positioned. Think of a busy doorway in a building. The entire building's ventilation can be affected if the doorway is blocked, even if the hallways elsewhere are clear. In ocean terms, deep convection is not a uniform process across the whole North Atlantic. It is patchy, seasonal, and sensitive to local conditions. A buoyancy cap that sits in the wrong place, at the wrong time, can reduce the depth of mixing that winter, and that one winter can influence the ocean's memory for years.

A historical anecdote makes this sensitivity more tangible. In the late 1980s and early 1990s, oceanographers began to measure an event that became famous in the climate literature: the Great Ocean Conveyor's "pulse", a period when deep convection in the Labrador Sea was unusually strong. The winter of 1989–1990, for example, produced deep mixed layers that helped form a thick layer of Labrador Sea Water, a relatively fresh but dense water mass that spreads through the North Atlantic at intermediate depths. That episode is remembered partly because it was so observable: repeated hydrographic surveys and later float measurements could trace the newly formed water. Yet the same region has also experienced winters when deep convection weakened or became shallow, sometimes for multiple years in a row. The contrast—vigorous mixing in some years, a muted response in others—underscores how close the Labrador Sea can be to the boundary between "open" and "shut."

The buoyancy cap mechanism offers a clean explanation for why that boundary can be crossed. Winter storms try to do two jobs at once. They cool the surface and they stir it. Cooling removes buoyancy; stirring tries to distribute the cooling downward by mixing. Both jobs can be thwarted by a strong near-surface density gradient. If the surface layer is fresh enough, even strong winds will mostly mix within the fresh layer itself. The cold produced at the surface does not reach the deeper, saltier water below. The deeper water, holding heat, remains out of reach. The storm may even make the cap more effective by spreading it laterally, sealing a larger area.

One can watch this "lid" forming in miniature at the mouth of a glacial fjord. Many Greenland fjords release meltwater into the sea as a surface plume. The plume is buoyant, so it spreads outward as a thin lens. Beneath it, seawater may be relatively warm, because Atlantic-origin waters can intrude into fjords at depth. That warm water melts glacier fronts underwater, producing more freshwater that rises and strengthens the plume. The fjord becomes a factory for stratification: freshwater at the top, warmer saltier water below, separated by a sharp gradient. Even without knowing the equations, it is easy to see why the top layer does not readily mix downward: it would require work to push light water under heavy water.

Farther from the coast, the geometry becomes less obvious but the physics is the same. Freshwater can be guided along the continental shelf by currents and winds. It can be stored temporarily in coastal boundary currents and then injected into the open ocean by eddies and storms. It can accumulate in the surface layer over a broad area, particularly if the winds do not favor deep mixing. In the Labrador Sea, for instance, winds can drive shelf waters into the basin's interior. That is not a poetic idea; it is an engineered pathway, a physical conveyor that takes low-salinity water that would otherwise hug the boundary and deposits it where winter convection would prefer a clear runway.

The word *halocline* comes up often in this context. It means a layer where salinity changes rapidly with depth. In a strongly stratified state near Greenland, the halocline can sit quite shallow. Above it is a cool, fresh layer; below it, warmer and saltier waters can lurk. This arrangement can feel counterintuitive, because people expect "warm water on top." In the high-latitude North Atlantic, the opposite can occur: warm water is trapped beneath a colder surface because salinity, not temperature, is providing the density contrast. The ocean can therefore store heat below a cold skin. That stored heat is not harmless. It can melt sea ice from below and, in fjords and along the shelf, it can reach glacier fronts and enhance melting. The lid does not merely prevent heat from escaping; it can help keep heat in the wrong place.

The practical measure of how difficult it is to mix a water column is often expressed through *stability* or *stratification strength.* Without drowning in formalism, the point is simple: stronger density differences with depth require more energy to mix away. The wind supplies energy, but the wind's effort is not unlimited. If the density jump is sharp enough, the wind will churn the surface into a rough, cold soup while the deeper ocean sits calmly below, like a room separated by a locked door.

This is where Greenland meltwater becomes climatically potent. Meltwater does not need to dominate the total freshwater budget of the North Atlantic to have an effect. It needs to be concentrated where it can intercept the ocean—atmosphere exchange that drives winter buoyancy loss. Greenland sits at the crossroads of currents and weather systems. Its surrounding seas are a meeting place for cold polar air outbreaks, powerful storms, and warm Atlantic inflows. The same geography that makes deep convection possible also makes it vulnerable: a cap placed there is placed in a strategic location.

A second concrete example brings the lid out of the realm of principle and into recent observation. In the central Labrador Sea, satellites have observed periods of unusually high sea surface height, a signature consistent with changes in the density structure of the water column: warmer and/or fresher upper layers expand, lifting the sea surface. At the same time, float and ship measurements have indicated concurrent warming and freshening in the upper ocean and a shoaling or cessation of deep convection over several years. The causal chain is not a single link—winds, heat loss, and remote inflows all play roles—but the pattern matches the lid mechanism: more buoyant surface waters, weaker vertical mixing, and a deeper ocean less directly ventilated by winter.

The "lid" language can mislead if it suggests an unbroken plastic cover. In reality, stratification can be punctured. Eddies peel off freshwater lenses and stir them. Storms create brief episodes of deep mixing. Sea ice formation can locally increase salinity by rejecting brine, briefly helping the surface grow denser. Yet the lid can still dominate the seasonal outcome. A winter's deepest mixing often depends on a few intense events—strong storms, cold-air outbreaks—that occur when the preconditioning is right. If the surface has been freshened enough, those events may no longer reach the depths they once did, even if they remain dramatic at the surface.

The human relevance is not confined to abstract circulation. A stratified, capped ocean changes how heat is shared between sea and air. If less ocean heat escapes in winter, the atmosphere above may be colder than it would otherwise be, even as heat builds below. The distribution of sea ice can shift. Marine ecosystems feel the difference because vertical mixing is how nutrients reach the sunlit surface. A capped ocean can become biologically quieter at the surface, not because life disappears, but because the seasonal replenishment of nutrients is weakened. Meanwhile, coastal Greenland communities and infrastructure contend with more meltwater and changing fjord conditions, and the global coastal population contends with the meltwater's contribution to sea level rise.

The essential physics of the lid can be conveyed with a single contrast: the same winter atmosphere acting over two different vertical structures. In one, surface waters are salty enough that cooling makes them heavy; they sink, dragging surface cold down and bringing subsurface water up, ventilating the ocean's interior. In the other, a fresh layer floats at the top; cooling may chill it to the freezing point, but it remains too light to sink. The deeper ocean retains its heat. The system looks superficially similar—a cold, stormy sea—but its internal exchange is transformed.

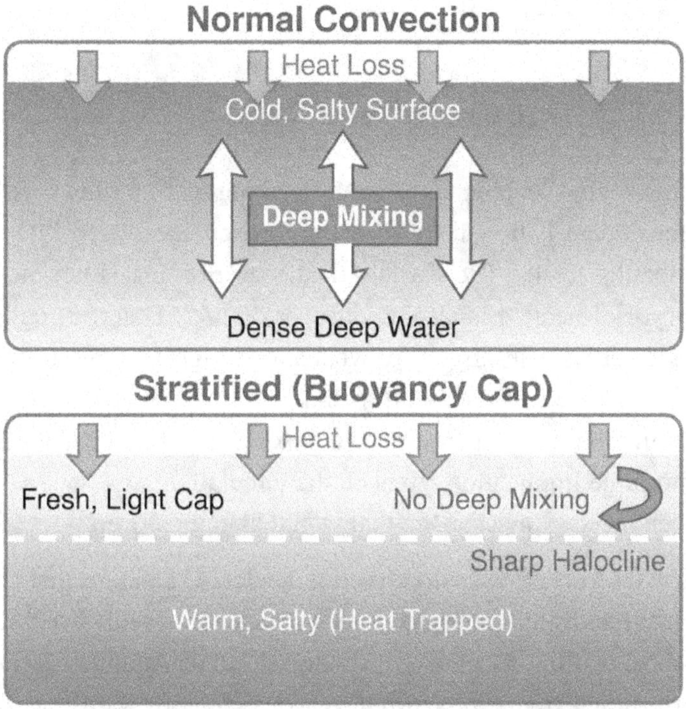

Comparison: Deep convection vs. Stratification.
Color intensity: Water density (darker = denser).

A final twist makes the lid especially unsettling: it can coexist with a surface that looks wintry and harsh. A person on a ship in February might see freezing spray, whitecaps, and perhaps sea ice not far away, and assume that the ocean must be vigorously mixing and cooling to depth. Yet the physics cares about the density ladder beneath the hull. If the ladder is missing a rung—if a buoyant freshwater layer interrupts the path—then the violence at the surface can be largely theatrical, churning a thin layer while deeper waters remain comparatively untouched. The ocean can be both storm-tossed and insulated, simultaneously losing heat from a skin and storing heat beneath it, held in place by a lid made of freshwater that began its life as snow on Greenland.

4.3 The Salt Feedback Trap

Freshwater has an obvious talent for making seawater float. The less obvious talent is how it can recruit the circulation itself into doing the floating for it. The North Atlantic overturning is not merely a conveyor of heat; it is also a conveyor of salt. That salt, delivered from lower latitudes, is part of what allows high-latitude surface waters to become dense enough to sink when winter strips away heat. A circulation that imports its own sinking ingredient carries an uncomfortable implication: weaken the circulation, and you may also weaken the very mechanism that would help it recover.

This is the core of the salt-advection feedback, a phrase that sounds technical until it is translated into ordinary terms. *Advection* means transport by flow, like a river carrying silt. The Atlantic Meridional Overturning Circulation (AMOC) carries salt northward near the surface as part of the broader northward flow of warm subtropical waters. That salt does not remain at the surface; it is mixed, transformed by air–sea fluxes, and eventually some of it is locked into dense waters that sink and return southward at depth. The feedback arises because the sinking needs density, density needs salt (especially near freezing), and salt arrives by a flow that depends on sinking.

The trap is not that the ocean is fragile in the everyday sense. For most of recent history it has been remarkably resilient, varying from year to year without falling apart. The trap is that the governing relationships are *nonlinear*: small changes can have small effects in some regimes and large effects in others. It is the difference between pushing a heavy door that is already swinging freely and pushing one that is just catching on the latch. In one case the same shove makes little difference; in the other it decides whether the door opens at all.

This nonlinear behavior is not a modern invention. Henry Stommel, an oceanographer with a gift for simple ideas, wrote down a toy model in 1961 that distilled the essence of thermohaline circulation into two boxes connected by a flow. One box represented low latitudes, where water is warmed and becomes light. The other represented high latitudes, where water is cooled and tends to become heavy. Salt was included as a second ingredient, because evaporation and precipitation move salt around indirectly by adding or removing freshwater. The striking outcome was not a particular number for flow strength, but the existence of two stable modes under the same external conditions: one with vigorous overturning and one with weak overturning. The system could sit in either mode, depending on its history. Once pushed far enough, it could switch abruptly.

"Box model" does not pretend to capture Labrador Sea eddies or the maze of Greenland fjords. Its power is that it makes the logic visible. In Stommel's picture, the high-latitude box becomes dense and draws in water from the low-latitude box. If an extra freshwater input lightens the high-latitude box, the flow weakens. That weakening reduces the import of salty water, making the high-latitude box even fresher and lighter. The same perturbation that started as a nudge becomes a self-reinforcing loop. In the model, there can be a threshold beyond which the "on" state cannot be maintained, and the system falls toward the "off" state. To return to the "on" state may require not merely removing the original perturbation, but reversing it strongly enough to cross a different threshold. This asymmetry is called *hysteresis*, a word meaning that the present depends on the past.

The Atlantic is not a pair of boxes. It is a rotating, stratified fluid full of turbulence and geography. Still, the salt feedback remains one of the most influential pieces of physical intuition for why the AMOC might have tipping behavior in some climate models. It also explains why freshwater inputs do not simply "dilute" the ocean in a passive way. They can alter the circulation that distributes salt, changing where salt is delivered and how effectively it counters buoyancy in the places where sinking is possible.

A concrete way to feel the loop is to imagine the upper North Atlantic as living on borrowed salt. The subtropics are relatively salty because evaporation there is strong: the sun and wind lift freshwater into the atmosphere and leave salt behind. Currents then move some of that salty water northward. High latitudes, by contrast, receive more precipitation and freshwater runoff, and the surface is frequently freshened by ice melt. Winter cooling tries to densify the surface, but if the surface has been excessively freshened, cooling may not be enough to make it sink. When sinking weakens, the engine that draws salty water northward also weakens. The high latitudes then lose part of their salt subsidy and become even more buoyant. The ocean is, in effect, cutting its own supply line.

There is a second twist that makes the trap harder to escape. Freshwater does not only change salinity; it changes how heat is stored. When a buoyant cap reduces vertical mixing, heat can accumulate below the surface. That subsurface warmth can also reduce density, further discouraging deep convection. In some model experiments where freshwater is artificially added to the North Atlantic (the so-called "hosing" experiments), a portion of the AMOC weakening is linked not just to surface freshening but to subsurface warming that reduces density where deep waters would otherwise form. Buoyancy is a combined effect of temperature and salinity, and the feedback loop can enlist both.

Real ocean observations offer hints of this coupled behavior. The early years of continuous AMOC monitoring at 26°N, begun in 2004, revealed substantial variability and a notable weakening during 2008–2012 compared with 2004–2008, associated with changes in the large-scale pressure and density structure of the Atlantic. Those measurements do not by themselves prove a salt-feedback trap; the ocean varies for many reasons. Yet they anchor the idea that the AMOC can shift significantly over a few years and that those shifts are connected to basin-scale density patterns. If the high-latitude North Atlantic is simultaneously being freshened and warmed at the surface, the ingredients for density reduction are being supplied on multiple fronts.

The salt feedback also clarifies why some freshening events feel more consequential than their size might suggest. The Great Salinity Anomaly that circulated through the subpolar North Atlantic in the late twentieth century was not simply an oddity of taste; it was a basin-scale buoyancy perturbation that moved through regions connected to deep convection and overturning. Later, in the 2010s, the eastern subpolar North Atlantic experienced an exceptional freshening in the upper ocean that drew attention precisely because it sat in a region usually buffered by salty inflow. In both cases, what matters is not only the immediate buoyancy change but how that buoyancy change influences the routing of salt and heat thereafter. A perturbation that reduces salt import can persist longer than the original freshwater pulse, because it reorganizes the circulation that would otherwise wash the perturbation away.

A powerful feature of the salt feedback is that it turns a spatial problem into a dynamical one. The Atlantic's salinity field is not uniform; it is patterned by the balance between evaporation and precipitation, by river inputs, by ice processes, and by circulation pathways. The AMOC is one of the processes that maintains this pattern, importing salty water northward. If the AMOC weakens, the salinity pattern can shift toward a fresher high-latitude state even without any additional freshwater input, simply because the salt transport is reduced. The feedback is thus a mechanism for persistence and amplification.

Oceanographers have tried to turn this intuitive loop into a measurable diagnostic. One influential idea focuses on the sign and magnitude of the *overturning freshwater transport* at the southern entrance to the Atlantic, around 34°S. Often denoted F_{ovS}, it is a bookkeeping measure of how the overturning circulation carries freshwater (or, equivalently, salt) into or out of the Atlantic basin. The concept is subtle but the logic is straightforward. If the overturning imports salt into the Atlantic (exports freshwater), then weakening the overturning reduces salt import, which tends to freshen the Atlantic and can promote further weakening: a positive feedback consistent with bistability. If, on the other hand, the overturning exports salt (imports freshwater), weakening it could have the opposite effect, potentially stabilizing the system.

In observations and reconstructions, F_{ovS} has often been estimated to be negative, interpreted by many as consistent with a regime where the salt-advection feedback could be active. Yet this diagnostic is not a crystal ball. Complex models include wind-driven gyre circulations, mixing processes, and biases in salinity and currents that can change the sign and meaning of F_{ovS}. Many climate models in intercomparison projects have historically produced the "wrong" sign compared with observational estimates, suggesting an AMOC that is too stable because it does not rely on importing salt in quite the right way. Even when the sign matches, the causal interpretation can be complicated: gyre transports and wind stress can rearrange salinity in ways that mimic or mask the overturning contribution. The diagnostic is therefore best treated as a useful clue about the system's background stability, not a definitive predictor of a near-term collapse.

The tension between simple feedback logic and messy reality is not a failure; it is a reminder of what makes the AMOC such a consequential subject. The circulation is not a single pipe with a valve.

It is an emergent pattern arising from winds, heat loss, freshwater inputs, mixing, and the constraints of continents. A feedback loop can be real and still be modulated, delayed, or redirected by other processes. A lid of meltwater may suppress mixing locally, but an unusually stormy winter could still deepen the mixed layer in patches. A fresh anomaly may enter the Labrador Sea, but its impact depends on whether it stays in the convective interior or is swept along the boundary. The salt feedback is a trap not because it is unavoidable, but because it can make recovery harder once the system has been pushed into a low-salt, low-density configuration.

One way to make the trap intuitive is to think in terms of *preconditioning*. Deep convection in the North Atlantic is not decided on the coldest day of winter; it is decided by how the water column has been set up over months. If autumn winds and currents have left the surface unusually fresh, the winter atmosphere must do extra work to remove buoyancy. If, at the same time, the northward delivery of salty subtropical water is weaker, the fresh anomaly is not easily erased. Each year's state becomes part of the starting point for the next. The ocean develops a kind of seasonal memory, and the salt feedback is one way that memory can become sticky.

At the dinner-table level, the story reduces to a paradox with teeth. The AMOC helps deliver the salt that makes northern waters heavy enough to sink. Freshening and warming make those waters lighter and reduce sinking. Reduced sinking weakens the AMOC. A weaker AMOC delivers less salt northward. That, in turn, makes the waters lighter still. The loop closes, and once it closes tightly enough, the system can become reluctant to reopen.

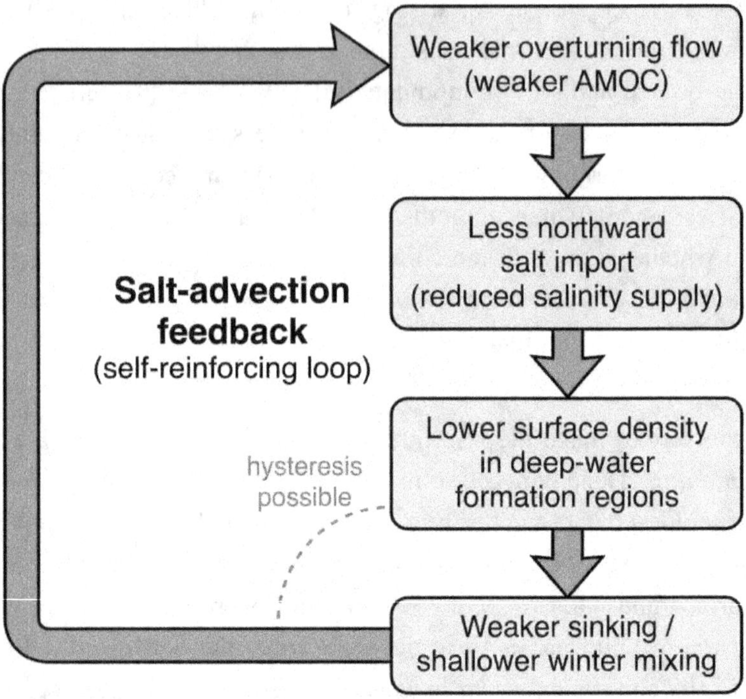

A freshwater pulse (from precipitation, rivers, or Greenland melt) can start the loop by lowering salinity and density at the surface.

The idea of a "point of no return" is often invoked too casually in climate conversations, as though the ocean were waiting for a single dramatic moment. The more honest picture is less cinematic and more unsettling: a gradual strengthening of buoyancy at the surface, winters that mix a little less deep, a salt supply line that becomes a little weaker, and a circulation that becomes a little less able to rebuild the conditions that once sustained it. The trap, once sprung, may not announce itself with a bang. It may announce itself with a quiet persistence: the return of spring arrives, the storms ease, and the ocean remains layered, holding onto its heat and its buoyancy as though it has learned a new habit.

Chapter 5

Ghosts of Climates Past

Civilization arose during a rare window of climatic stability, but the geologic record reveals a much more volatile history beneath our feet. By interrogating the chemistry of ancient ice and ocean sediments, we discover that the ocean's heat conveyor is not a steady flow, but a temperamental system prone to violent, rapid shutdowns.

5.1 Decoding the Deep Archive

No one has ever watched the Atlantic overturning circulation switch gears in real time over ten thousand years. The instruments that measure currents with exquisite precision today—moorings, floats, satellites—have existed for decades, not millennia. Yet the ocean has been moving heat and salt around the planet for as long as there has been an ocean, and its older behavior still matters: it tells us what kinds of reorganizations are physically possible, how fast they can unfold, and what fingerprints they leave behind.

The way into that deep history is indirect. Climate scientists use *proxies*: natural recorders that respond to temperature, salinity, ice volume, wind, and biological productivity, and that preserve those responses in ice, mud, and skeletons. Proxies are not photographs of the past; they are coded messages. Decoding them takes a blend of chemistry, physics, and an almost forensic attention to how a signal could have been altered after it was written.

Two archives dominate the story of ancient ocean circulation. One is polar ice, laid down snowflake by snowflake, compacted into layers that can be counted like tree rings in places where annual layering survives. The other is seabed sediment, falling steadily through the water column and mixing with bursts of debris when icebergs melt or rivers flood. Ice cores give an atmospheric and high-latitude view; marine sediments give an ocean view. The power comes from linking them.

A useful starting point is density, because density is the currency the overturning circulation trades in. Seawater becomes dense by cooling and by becoming salty. Cooling is straightforward: colder water is heavier. Salinity is trickier, because salt is added and removed by processes at the surface—evaporation concentrates salt; rain and river runoff dilute it; sea ice formation leaves salt behind; sea ice melting adds fresh water back. In the North Atlantic, the overturning circulation depends on surface waters becoming dense enough to sink and feed the deep return flow. If something prevents that sinking—by warming the surface or freshening it—the circulation weakens.

Proxies rarely measure density directly. Instead they measure temperature, salinity, ice volume, ventilation, or the transport of particular water masses. Each of those can be tied back to overturning strength, but only with careful reasoning and, ideally, multiple lines of evidence.

Ice cores begin as snowfall. As each storm passes over Greenland or Antarctica, water vapor condenses and falls as snow. The key is that water molecules are not all identical. Some contain the common oxygen isotope, ^{16}O; a small fraction contains the heavier ^{18}O. Because the heavier molecule behaves slightly differently during evaporation and condensation, the ratio of ^{18}O to ^{16}O in snowfall carries information about the conditions under which that snow formed. The ratio is usually reported as $\delta^{18}O$, a small deviation (in parts per thousand) relative to a standard.

To an excellent first approximation, Greenland $\delta^{18}O$ becomes less negative ("heavier") when local temperatures are warmer, and more negative ("lighter") when they are colder. That is why the Greenland ice cores read like a jagged temperature history, sometimes changing with alarming speed. But $\delta^{18}O$ is not a thermometer in the everyday sense. It is a tracer of the water cycle: where moisture came from, how cold the air masses were, how storms tracked, and how precipitation formed. Those complications can be nuisances, but they are also opportunities: they mean the ice records atmospheric circulation changes as well as local temperature.

Ice cores also trap ancient air. As snow compacts into firn—a porous, crunchy stage between snow and solid ice—the pathways between grains gradually close. Air is sealed into bubbles. Those bubbles contain past concentrations of greenhouse gases like CO_2 and CH_4, and they allow direct comparisons between temperature proxies and atmospheric composition. For ocean circulation, the bubbles matter in another way: methane changes can be tied to shifts in tropical and boreal wetlands, and those shifts can be synchronized with abrupt climate events that are often connected to Atlantic circulation changes. The ice also contains chemical species deposited from sea spray, dust, volcanic eruptions, and even biomass burning, each adding context about winds and aridity.

A historical anecdote captures the emotional jolt of this archive. In the early 1990s, when deep Greenland cores such as GRIP and GISP2 were being extracted and analyzed, the scientific community expected glacial climates to be cold and variable, but the sharpness of the shifts was still startling. The isotopic profile did not show gentle, rounded transitions. It showed steps. Some warmings looked, in the ice record, as though Greenland had changed its winter coat within a human lifetime. That realization helped turn the phrase "abrupt climate change" from a curiosity into a research program, and it forced oceanographers to take seriously the idea that the North Atlantic could reorganize rapidly.

Yet ice cores, by themselves, do not measure ocean currents. They tell us about the atmosphere and high-latitude climate, and they do it with exceptional temporal precision. To translate that into ocean circulation, marine sediments are essential.

Seabed sediments accumulate in layers, but with more smudging than ice. Tiny organisms, currents, burrowing animals, and chemical reactions can mix or alter the upper centimeters. Even so, sediments hold an astonishing inventory. They contain microscopic shells of plankton that once lived near the surface, and of bottom-dwelling organisms that lived on or within the mud. They contain dust grains blown from continents, and occasional coarse fragments dropped from melting icebergs. They contain chemical tracers adsorbed onto particles or embedded within minerals. Each of these components can be dated and interpreted.

Among the most important witnesses are foraminifera, or forams: single-celled protists that build calcium carbonate shells. When a planktonic foram grows in the surface ocean, it incorporates oxygen and carbon from seawater into its shell. The same isotopic logic applies as in ice, but with a twist. The δ^{18}O of a shell depends on two things: temperature and the δ^{18}O of seawater. Seawater δ^{18}O, in turn, depends on salinity and on global ice volume. During glacial periods, large ice sheets store a great deal of the light isotope (^{16}O), leaving the ocean enriched in ^{18}O. So a higher (heavier) foram δ^{18}O can mean colder water, saltier water, more global ice, or some combination.

That ambiguity sounds discouraging until one sees how it becomes a strategy. If you have an independent estimate of temperature, you can back out seawater $\delta^{18}O$ and infer changes related to salinity and ice volume. If you have paired measurements from surface-dwelling and deep-dwelling forams, you can reconstruct gradients—how different the surface was from the deep—which are intimately connected to stratification and overturning.

Temperature itself can be inferred from more than one proxy. The magnesium-to-calcium ratio (Mg/Ca) in some foram shells increases with temperature, providing an independent thermometer. Certain organic molecules produced by marine microorganisms change their structure depending on growth temperature, leaving another set of paleothermometers in sediments. Each method has caveats: biology matters, dissolution matters, habitat depth matters. But when multiple thermometers agree on the direction and approximate size of a change, confidence grows.

The deep ocean leaves its signatures too. Bottom water temperature can be reconstructed from benthic forams, while the ventilation of deep waters—how recently a water parcel was at the surface—can be estimated using radiocarbon. A well-ventilated deep Atlantic suggests vigorous formation of deep water in the North Atlantic. An old, poorly ventilated deep Atlantic suggests reduced renewal, whether because formation slowed, shifted location, or was blocked by surface freshening.

Dating is the quiet engine under all this work. A proxy without a reliable clock is a rumor. Ice cores can sometimes be counted year by year; they also contain volcanic ash layers and other markers that can be tied to known events. Marine sediments are harder. Radiocarbon dating applies to the last $\sim$ 50,000 years and must be corrected for the marine reservoir effect—the fact that surface waters can be depleted in ^{14}C compared to the atmosphere, making them appear artificially old. Beyond radiocarbon, scientists use oxygen isotope stratigraphy, matching global ice-volume signals; they use geomagnetic excursions; they use tephra layers (volcanic ash) when available; and they use tuning approaches that align sediment properties with well-dated orbital variations. Each method introduces uncertainty. A common, practical skill in paleoclimate science is knowing when a timing offset is meaningful and when it could be a dating artifact.

Even with careful dating, there is a second problem: the ocean is a three-dimensional, moving fluid, and sediments are point measurements. A single core taken from a single site may tell a beautiful story about that site, but overturning circulation is a basin-scale phenomenon. The solution is networks. When many cores from different locations show coordinated changes, a circulation interpretation becomes far more compelling.

How does one go from "temperature changed" to "overturning changed"? Often by looking for patterns that are hard to explain without moving water masses.

Consider the North Atlantic as a stage with distinct actors: warm, salty subtropical waters flowing north near the surface; cooler, fresher waters nearer the Arctic; and deep waters formed in the subpolar North Atlantic and Nordic Seas. When deep water formation is strong, the deep Atlantic is filled with relatively young, oxygen-rich waters with characteristic chemical properties, and the boundary between northern- and southern-sourced deep waters sits at certain depths. When formation weakens, southern-sourced waters can encroach northward and upward, and the deep Atlantic becomes older in radiocarbon age.

A concrete example comes from the use of radiogenic neodymium isotopes (ε_{Nd}) as a water-mass tracer. Different ocean basins and continental margins impart distinct neodymium signatures to seawater. While the full chemistry is intricate, the practical payoff is intuitive: a shift in ε_{Nd} recorded by sediments at a given depth can indicate a changing mixture of northern versus southern deep waters. That is not a direct speedometer for the overturning circulation, but it is a strong indicator of reorganized deep-water geometry—the plumbing layout.

For a more direct connection to export of deep water, oceanographers have turned to a clever pair of naturally occurring radionuclides: ^{231}Pa and ^{230}Th. Both are produced in seawater by radioactive decay of uranium. Thorium is extremely particle-reactive and is removed quickly to sediments near where it is produced. Protactinium is less particle-reactive and stays dissolved longer, making it more sensitive to transport by currents before being removed. In the modern Atlantic, a vigorous overturning circulation exports dissolved ^{231}Pa southward, lowering the amount that settles locally relative to ^{230}Th. When export weakens, more ^{231}Pa is left behind and the sediment ratio $^{231}\mathrm{Pa}/^{230}\mathrm{Th}$ rises. This approach has been central to arguments that the Atlantic overturning circulation weakened strongly during certain cold intervals of the last deglaciation.

But proxies that seem almost tailor-made for overturning still require humility. Particle fluxes can change. Boundary scavenging along margins can trap protactinium. Circulation pathways can shift so that a site no longer sits under the same "conveyor belt." These are not fatal flaws; they are reminders that the ocean is clever at hiding its motives. The most convincing reconstructions are those where multiple proxies—water-mass tracers, ventilation ages, temperature and salinity indicators—tell compatible stories.

One of the most vivid sedimentary markers of abrupt North Atlantic change is ice-rafted debris: coarse grains and pebbles that could not have drifted to the deep ocean by ordinary currents. They arrived frozen into icebergs, released as the ice melted. In cores taken from the North Atlantic, layers rich in such debris appear as geological confetti, proof of massive iceberg discharge events. These layers matter for overturning because iceberg armadas imply freshwater delivery to the surface ocean. Freshwater makes surface water lighter, and lighter surface water resists sinking. The physical link between freshwater and reduced deep water formation is direct, a rare gift in paleoclimate interpretation.

Here, the combination of archives becomes especially powerful. If a debris-rich layer in marine sediments can be aligned in time with a sharp cooling in Greenland ice and with changes in deep-water tracers, the case for an ocean-driven climate event strengthens. Bond and Broecker (1993) were influential in showing tight correlations between North Atlantic sediment signals and Greenland ice-core variability, helping to cement the idea that the ocean was not merely reacting to atmospheric shifts but participating as an active, sometimes dominant, agent.

Yet an important subtlety remains: the overturning circulation is not just a north—south pipe; it is a system of currents with recirculations, eddies, and multiple formation regions. "Strength" can mean different things: the volume transport of deep water, the heat transport to high latitudes, or the extent of northern-sourced waters filling the deep basin. Proxies can target different aspects. A water-mass tracer might show a shift in geometry with only modest change in total transport. A heat transport change might be large even if volume transport changes modestly, if the temperature contrast changes. This is why paleoclimate debates can persist even when all parties agree that something dramatic happened.

A good proxy story includes how the signal might have been distorted. In marine sediments, one common distortion is bioturbation—mixing by organisms—which can smear a sharp event into a gradual one. Dissolution of calcium carbonate on the seafloor can selectively remove more fragile shells, altering species composition and potentially biasing isotope measurements. Foraminifera do not live everywhere in the same season or depth; if a species prefers summer surface waters, it records summer conditions more than annual means. In ice cores, diffusion in firn can smooth some chemical signals, and the air in bubbles is typically younger than the surrounding ice because it is trapped after the snow has been buried for some time. None of these complications make the archives unusable; they simply mean every sharp line on a graph has a story of its own.

An underappreciated practical detail is that proxy science often advances when someone finds a way to measure something that used to be invisible. The first measurements of stable isotopes in polar precipitation were laboratory feats. The extraction of ancient air without contamination required careful engineering. The ability to measure trace metal ratios like Mg/Ca in tiny shells depended on improvements in mass spectrometry and sample cleaning protocols. Each technical leap widened the set of questions that could be asked about the past ocean.

At the end of all this decoding sits a picture that is never drawn from a single piece of evidence. The Greenland ice tells of sudden cold snaps and abrupt warmings, paced not only by slow changes in Earth's orbit but by internal reorganizations. The sediments tell of shifting water masses, changing deep ventilation, and episodes when the surface North Atlantic was freshened by meltwater and icebergs. Chemical isotopes act as translators between the two, turning ancient temperatures and salinities into inferences about whether surface waters were sinking efficiently, whether deep waters were being renewed, and whether the ocean's great northward heat transport faltered.

It is easy to miss what a peculiar achievement this is. The ocean erases its tracks as it moves; yesterday's currents are overwritten by today's. Yet, scattered across the polar ice and pressed into the seabed, the ocean has left behind enough clues to reconstruct not just that it changed, but how it changed: where the surface waters freshened, where the deep ocean aged, where the temperature gradients steepened, and where the fingerprints of distant water masses appeared.

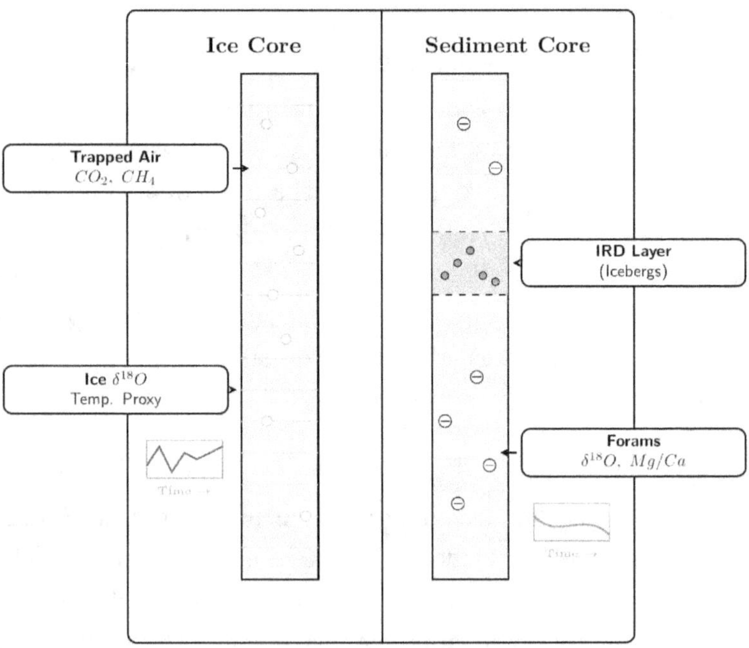

The deep archive does not hand over the past ocean as a single number. It offers a set of constraints: surface waters warmed or cooled; they freshened or grew saltier; deep waters became younger or older; northern- or southern-sourced waters expanded; iceberg discharge rose or fell. From those constraints, an image of circulation emerges that is more persuasive than any one measurement could be. It is a portrait drawn from chemistry and mud and ancient air, showing that the ocean's great overturning is not an unchanging backdrop but a dynamic player whose older moods are still legible, if one learns to read the code.

5.2 The Great Shutdown

Imagine standing on the shores of the North Atlantic at the edge of the last ice age, at a moment when the planet is *already* warming. Glaciers are retreating, landscapes are draining, coastlines are shifting. Then, within a span short enough to fit inside a family saga, winter seems to return with a vengeance. In Europe, the warming stalls; in Greenland, the cold bites hard. A world that had begun to loosen its grip on ice suddenly clenches again.

That jarring reversal is the Younger Dryas, beginning about 12,900 years before present and ending around 11,700 years before present. It is not the only such interruption, but it is the most famous because it sits right at the doorway to the relatively stable climate in which agriculture, cities, and everything that followed took shape. In the language of ocean physics, the Younger Dryas is a suspected episode of a weakened Atlantic Meridional Overturning Circulation (AMOC): the system that carries warm, salty surface waters northward and returns cold, dense waters southward at depth. In the language of daily life, it is what happens when the North Atlantic's heat delivery falters.

The core idea is almost disarmingly simple. The AMOC depends on seawater becoming dense enough to sink in a few crucial regions of the subpolar North Atlantic and the Nordic Seas. Density depends mainly on temperature and salinity. Cooling makes water denser; freshening makes it lighter. Pour enough fresh water into the right place, and the surface ocean becomes a buoyant lid. Convection weakens or shuts down, deep water formation slows, and the overturning circulation slackens. Less warm water arrives from the south. The North Atlantic cools. Winds, sea ice, and storms rearrange themselves in response. A regional shift becomes a hemispheric story.

What makes the Younger Dryas and the Heinrich events so unsettling is not only their magnitude, but their timing. They are interruptions and shocks embedded within a broader, slower deglaciation. The climate system is not behaving like a smoothly warming planet; it is behaving like a system with thresholds.

Long before the oceanographic term AMOC became common, geologists were noticing strange layers in North Atlantic sediments: horizons rich in coarse grains, sand, and small stones that do not belong in the middle of a deep-ocean mud sequence. The explanation is beautifully mundane. Those grains were carried out to sea frozen into icebergs. When the icebergs melted, the debris fell straight down, sprinkling the seabed. These are the Heinrich layers, named after the marine geologist Hartmut Heinrich, who in the 1980s documented a series of such layers in sediment cores and argued that they represented massive, episodic discharges of ice from the Laurentide Ice Sheet into the North Atlantic.

It is hard to overstate what is implied by a Heinrich layer. An iceberg is not just frozen water; it is a delivery vehicle for fresh water *and* for a geographical fingerprint. The mineral grains and rock fragments can often be traced back to source regions, especially around Hudson Strait, pointing to a particular sector of the Laurentide Ice Sheet as the culprit. A Heinrich event is, in effect, a temporary re-plumbing of a continent-scale ice sheet, sending a burst of fresh water and ice into waters that are normally saltier and primed for sinking.

The ocean does not respond politely. Fresh water spreads over the surface, lowering salinity. Sea ice may expand, and when sea ice forms it rejects salt into the underlying water, which can *locally* increase density, but it also insulates the ocean from the atmosphere and changes where heat is lost. Meanwhile, colder surface conditions reduce evaporation, which can further reduce the salinity-building effect of the atmosphere. The feedbacks are interlocking: a freshening lid discourages sinking, weaker sinking reduces northward heat transport, colder conditions encourage more sea ice, and expanded sea ice changes both buoyancy forcing and atmospheric circulation. None of these steps requires exotic physics. The drama comes from their alignment.

The Younger Dryas is often introduced as a freshwater story with a famous protagonist: proglacial Lake Agassiz. As the Laurentide Ice Sheet retreated, it dammed meltwater into huge lakes along its margin. Lake Agassiz was among the largest, sprawling over parts of what are now Canada and the northern United States. It did not sit there quietly. Its outlets shifted as ice dams grew, failed, or were overtopped, and as the land rebounded after being relieved of the ice sheet's weight. When a large body of fresh water suddenly finds a new route to the ocean, the effect downstream can be dramatic.

The hypothesis that a re-routing or outburst of Lake Agassiz helped trigger the Younger Dryas is compelling because it ties a specific, physical event to a specific vulnerability of the ocean: the sensitivity of North Atlantic deep water formation to surface salinity. Yet the details remain an active area of research and debate, in part because meltwater routing is messy. Water can flow east toward the St.

Lawrence and the North Atlantic, or northward toward the Arctic, or south toward the Gulf of Mexico, depending on ice geometry and topography at a given moment. The Younger Dryas is not a controlled experiment; it is a complicated natural one.

A particularly elegant approach to this problem uses chemical "fingerprints" preserved in foraminifera living near key gateways. Carlson and Clark (2007), for example, used geochemical tracers to infer shifts in freshwater routing near the St. Lawrence region at the onset of the Younger Dryas, and they estimated a freshwater flux on the order of a few hundredths of a Sverdrup (a Sverdrup, abbreviated Sv, is a million cubic meters per second). That number is hard to visualize until one translates it into what it means for the upper ocean. Spread over months to years into the right region, it is enough to measurably freshen the surface and nudge convection toward failure. Small changes in salinity can matter enormously because the density difference driving sinking is itself a small difference between large numbers.

A concrete image helps. In the modern North Atlantic, winter storms strip heat from the ocean surface. In regions like the Labrador Sea, surface waters become cold and dense, and in some years they convect deeply, overturning and ventilating the interior ocean. Now imagine that the same winter winds arrive, but the surface is capped by a fresher layer. The winds can cool the surface, but cooling alone may not be enough to make the water heavy enough to sink through the fresh cap. The energy of the storms goes into stirring the lid rather than overturning the whole column. The ocean remains stratified. The deep ocean does not get its usual infusion of newly formed dense water. The conveyor slows, not because the winds stopped, but because the surface boundary condition changed.

At the heart of the "shutdown" claim is the question of how we know the AMOC weakened, and by how much. Temperatures and ice-rafted debris tell us the surface North Atlantic changed dramatically, but they do not directly measure overturning transport. One of the most influential tools for that purpose has been the sedimentary ratio of $^{231}Pa/^{230}Th$, a clever kinematic proxy rooted in how the ocean removes these isotopes. Because ^{230}Th is rapidly scavenged onto sinking particles, it mostly records local production. Because ^{231}Pa lingers longer, it can be carried by currents before being removed. A strong overturning circulation exports more ^{231}Pa out of the North Atlantic, lowering the local sediment ratio. A weakened overturning leaves more behind, raising the ratio.

McManus and colleagues (2004) applied this approach to North Atlantic sediments and argued for a near-collapse of overturning during Heinrich Stadial 1, roughly 17,500 years ago, and for a sharp weakening into the Younger Dryas. Their work was part of what made the "great shutdown" feel like more than a metaphor. It suggested that the ocean's deep engine could be throttled back drastically, and that it could resume relatively quickly when conditions changed.

Science, however, rarely grants a single decisive measurement. Later studies, including work by Bradtmiller and McManus (2014), emphasized nuance: the $^{231}Pa/^{230}Th$ signal can depend on where you measure it, on how particle fluxes and boundary scavenging behave, and on how the geometry of currents changes. The difference between "collapsed" and "strongly weakened" is not semantic. It changes how one imagines the ocean's stability and how close the system came to a truly different operating mode. A circulation that nearly stops is a different beast from a circulation that limps along.

That debate is not a weakness of the field; it is what careful inference looks like when the data are indirect. The ocean is not obliged to change in the simplest possible way. It can shift where deep waters form, alter pathways, or redistribute the balance between different branches of the overturning while still moving substantial amounts of water. The best reconstructions treat a shutdown as a family of related changes: reduced deep ventilation, altered water-mass distributions, colder surface conditions in key sinking regions, and evidence of freshwater forcing. When many of these lines point together, the case becomes difficult to dismiss.

A striking feature of these events is their uneven geography. A weakened AMOC does not cool the entire planet uniformly. It disproportionately cools the North Atlantic region because that is where the missing heat would have been delivered. Europe's climate is strongly shaped by prevailing winds and atmospheric circulation, so the link is not as simple as "Gulf Stream equals warm Europe," but a colder North Atlantic tends to shift storm tracks and sea ice, amplifying winter severity in parts of the region. Meanwhile, the Southern Hemisphere can behave differently. The "bipolar seesaw" idea, developed in part through the work of Stocker and Johnsen (2003), captures the observation that when the North Atlantic cools abruptly, Antarctica may warm more gradually, consistent with a redistribution of heat between hemispheres mediated by the ocean's deep reservoirs. The seesaw is not a literal lever; it is a statement about how ocean heat transport couples distant regions with characteristic time lags.

That global reach matters because it reframes the Younger Dryas and Heinrich events as more than local curiosities. A disturbance in one corner of the Atlantic can echo into monsoons, tropical rainfall patterns, and sea ice far away. The ocean is not a passive heat bath. It is an active organizer of climate patterns.

A historical anecdote, often overlooked, comes from the practical side of field science. Recovering a long, undisturbed sediment core from the North Atlantic is not like taking a sample from a quiet lake. Ships fight weather, currents, and equipment failures. Cores can be lost, smeared, or recovered in pieces. The triumph of those early Heinrich-layer identifications and later geochemical reconstructions is partly a triumph of stubbornness: going back to sea, improving coring technology, building stratigraphies that could be matched across basins. In paleoclimate work, the decisive insight often arrives only after the tenth core, when patterns finally become undeniable.

Those patterns reveal another unsettling aspect of the great shutdowns: they are not one-off freak accidents. Heinrich events recur during glacial periods, roughly on millennial timescales, suggesting an internal rhythm involving ice-sheet dynamics and ocean feedbacks. The Younger Dryas, though unique in its timing and context, fits into a broader tendency for the deglacial climate to lurch rather than glide.

It is tempting to imagine a single trigger: a catastrophic flood here, an iceberg armada there. But the emerging picture is more like a loaded system. Ice sheets were large and unstable; sea level was rising; freshwater reservoirs were enormous; winds and sea ice were shifting; atmospheric greenhouse gases were changing. A freshwater pulse might have been the final nudge, but the ocean's sensitivity—its readiness to respond sharply—was shaped by the entire deglacial background state.

The "freshwater lid" concept is easy to draw and, in essence, easy to understand. It is also easy to underestimate. Fresh water does not need to fill the Atlantic to matter. It needs to be delivered to the right places at the right times, especially where winter cooling normally creates dense waters. A thin layer can have an outsized effect because it changes the *vertical* exchange: the ability of the surface to communicate with the deep. Once that communication is disrupted, the deep ocean becomes insulated from atmospheric forcing, and the climate system's memory lengthens. The ocean can then store anomalies—of heat, of salinity, of carbon—and release them later when circulation resumes.

One can think of the AMOC as a heat pump that also shuffles salt. When it weakens, the North Atlantic receives less imported heat, but it also receives less imported salt. That second change can reinforce the first. Less salt imported means surface waters are more easily freshened by precipitation, runoff, or melting, which makes sinking even harder. This is the essence of the salt-advection feedback first made explicit in simplified form by Henry Stommel (1961): a circulation that imports salt to its sinking regions can, under some conditions, stabilize itself, but if that salt import is reduced, the system may slip toward a weaker state that is self-consistent. The great shutdowns are not merely about adding fresh water; they are about tipping a coupled heat-and-salt machine into a different mode of operation.

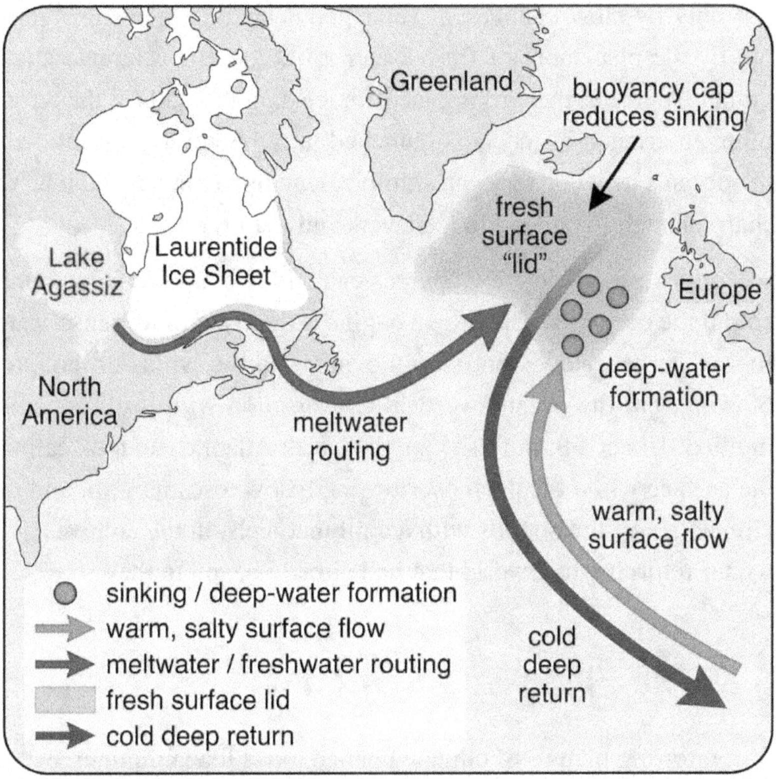

The map is schematic, but the physics is not. The subpolar North Atlantic is one of the few places where the surface ocean can readily communicate with the abyss. If the surface layer is made even slightly more buoyant, the whole system can hesitate. The hesitation is amplified because the AMOC, once weakened, tends to import less salt into the very regions that need it to sustain sinking. A freshwater pulse can therefore do double duty: it freshens directly, and it can push the circulation into a state where it freshens itself.

That logic makes the Younger Dryas and the Heinrich events feel less like ancient curiosities and more like case studies in sensitivity. They show a climate system that can be pushed off its preferred track not only by slow changes in solar geometry or greenhouse gases, but by the placement of fresh water at the ocean's strategic choke points. A lake in the wrong place, an ice dam that fails at the wrong time, an armada of icebergs launched into the wrong current—and the ocean's heat delivery can stumble, leaving behind a cold interval sharp enough to be read in ice layers and mud.

A world warmed by a few degrees can still host a sudden, stubborn cold if the ocean's salt-and-heat engine is deprived of its dense-water fuel. The water does not stop moving, and the winds do not stop blowing, but the ocean's vertical conversation with itself becomes muffled. Under a thin, fresh cap, the North Atlantic can look calm at the surface while the deep overturning below loses its grip, and the climate above it responds with a chill that feels, in the archives, like winter returning to a world that had already begun to thaw.

5.3 The Flickering Switch

A comforting picture of climate change looks like a dimmer switch: turn the knob slowly and the room brightens slowly. That picture is not wrong for many parts of the system. Ice sheets grow and melt with the patience of centuries; the deep ocean stores heat with the inertia of generations. But the Greenland ice tells a different story for large parts of the last glacial period. There, the climate behaves less like a dimmer and more like a faulty light switch: long stretches of gradual drift punctuated by sudden snaps.

These snaps are the Dansgaard–Oeschger cycles, usually shortened to D–O cycles after Willi Dansgaard and Hans Oeschger, who helped make Greenland ice-core chemistry a language for abrupt climate change. They are not subtle wiggles. In the Greenland isotope record, temperatures appear to jump by several degrees—sometimes by as much as about 10°C in the local signal—over intervals that can be as short as decades. Then, after the sudden warming, the system tends to slide back toward colder conditions over centuries to a millennium, until the next jump.

The shape matters. It is not a symmetric oscillation. It is a saw-tooth: slow cooling, abrupt warming. That asymmetry is a clue about mechanism. Climate has many ways to drift slowly: weak feedbacks, gradual forcing, the long tail of ocean and ice response. But abrupt warming implies a threshold was crossed, allowing the system to reorganize quickly into a different state.

It is tempting to frame D–O cycles as curiosities of an ice age world, irrelevant to the present. That would be a mistake. The modern climate is different, but the physics that allows abruptness is still there. If a system can switch quickly under one set of conditions, it is worth asking what prevents switching under another.

D–O cycles are often introduced through a plot: δ^{18}O in Greenland ice, a proxy for local temperature and atmospheric circulation, swinging up and down in repeated steps. The ice-core signal is not a perfect thermometer, but its abruptness is too large, too coherent across multiple cores, and too tightly correlated with North Atlantic marine changes to dismiss as a local quirk. When Greenland warms abruptly, the North Atlantic surface ocean tends to warm as well; when Greenland cools, the ocean cools and sea ice expands. The atmosphere, the ocean surface, and (with some lag) the deep ocean move together.

The key question is whether the ocean is merely following, or whether it is part of the switch mechanism. A useful way to ask that question is to look for a component of the climate system that can do two things at once: (i) store energy and salt over long enough periods to create slow drift, and (ii) reorganize rapidly when a threshold is crossed. The AMOC is a strong candidate because it is both a transporter and a gatekeeper. It transports heat northward, and it controls whether the surface North Atlantic can efficiently ventilate the deep ocean.

A slightly uncomfortable fact sits behind this: the AMOC is not guaranteed to have a single stable strength for a given external climate. In the simplest conceptual models, it can have more than one stable regime. Henry Stommel (1961) showed, in a deliberately stripped-down two-box model, that the same external forcing could admit two different circulation states: a stronger, "on" state and a weaker, "off" state, separated by a threshold. That result did not claim the real ocean is literally two boxes. It did something more important: it demonstrated that multi-stability is not a fantasy. It can arise from ordinary physics when the circulation moves salt in a way that feeds back on density.

The salt-advection feedback is easy to state. If the overturning circulation is strong, it can import more salty water into the North Atlantic sinking regions. More salt tends to make the water denser, which favors sinking and helps maintain a strong overturning. If the circulation weakens, less salt is imported; the surface becomes more easily freshened by precipitation, runoff, or melting; the water becomes less dense; sinking becomes harder; the overturning weakens further. A feedback loop appears, capable of amplifying modest perturbations into large changes.

A feedback loop does not automatically create a switch. It creates the *possibility* of one, because it allows the system to have preferred states that reinforce themselves. Add a threshold and hysteresis—the idea that the path into a state can differ from the path out of it—and the climate begins to resemble a flickering light near a loose electrical connection. Small nudges can toggle it back and forth.

One reason D–O cycles feel unnerving is that they violate a common intuition about large systems: big things should move slowly. The ocean is big. How can it change quickly?

The answer is that the ocean does not need to change everywhere at once to cause a big climate response. It needs to change at key choke points: places where a small shift in surface buoyancy can decide whether the winter mixed layer reaches a few hundred meters or plunges far deeper; places where sea ice can expand or retreat, changing not only albedo (how much sunlight is reflected) but also the exchange of heat and moisture between ocean and atmosphere. Once those choke points flip, the atmosphere reacts rapidly because air moves fast, carrying the new pattern of heat fluxes and temperature gradients into storm tracks, winds, and precipitation. The ocean then responds on multiple timescales: surface changes quickly; the deep ocean adjusts more slowly; the entire system relaxes until the next threshold crossing.

That logic suggests a simple narrative for the sawtooth shape. During the slow cooling phase, the North Atlantic may be gradually freshening or becoming more strongly stratified, perhaps due to background glacial conditions: expanded ice sheets, altered winds, changes in sea ice export, and a different hydrological cycle. The overturning circulation weakens or shifts, reducing northward heat transport. Greenland cools. Sea ice expands, which can further cool the region and reinforce stratification. The system creeps toward a threshold.

Then something pushes it over. The push might be small and need not be the same every time: a change in freshwater input, a shift in winds, a fluctuation in sea ice export through Fram Strait, or an internal instability of the coupled ocean—ice system. Once the surface becomes salty enough in the right region, or once sea ice retreats enough to allow strong winter heat loss, deep convection can restart vigorously. Dense water forms, the overturning strengthens, warm water is drawn northward, and Greenland warms abruptly. The warming is abrupt because convection and the atmosphere respond quickly once conditions permit.

After the jump, the slow drift resumes, now in the opposite direction, as feedbacks and background forcing gradually set up the next weakening. The sawtooth is not a metronome. It is the signature of a system repeatedly approaching, crossing, and retreating from a threshold.

It helps to see the geometry of the real Atlantic, because the AMOC is not a single pipeline but an intertwined set of currents. Warm waters flow northward near the surface and release heat to the atmosphere; colder waters return southward at depth. Deep water formation is concentrated in certain regions, notably the Labrador and Nordic Seas, but the details vary with climate state and with the balance between surface cooling and freshening. In glacial climates, sea ice can extend farther south, and the boundary between open ocean and ice can move dramatically from decade to decade. That boundary matters because it controls where the ocean can lose heat in winter. A slight change in sea ice extent can therefore reposition the main sites of deep convection, changing how efficiently the ocean ventilates its interior and how much heat it transports north.

This is where D–O cycles and Heinrich events begin to look like relatives rather than strangers. Heinrich events are marked by massive iceberg discharges and strong freshwater forcing. D–O cycles are smaller, more frequent, and can occur without a Heinrich layer. Yet they appear embedded in a broader glacial rhythm in which some sequences of D–O oscillations culminate in a Heinrich event. Bond and Broecker (1993) helped establish the close relationship between Greenland abrupt changes and North Atlantic sediment signals, strengthening the argument that the ocean is part of the switch. The picture that emerges is not of one mechanism operating alone, but of a coupled system: ice sheets, sea ice, ocean circulation, and atmosphere all interacting, sometimes producing rapid jumps and sometimes producing larger, iceberg-laden disruptions.

A concrete human-scale comparison makes the timescales vivid. A temperature shift in Greenland over a few decades is the difference between a grandparent's childhood winters and a grandchild's. It is short enough that ecosystems, ice margins, and human societies (had they been present in large numbers at those latitudes) would experience it as an abrupt upheaval rather than a slow trend. When the climate record contains repeated events like that, the natural question is not "Did this really happen?" but "What kind of system does this?"

The scientific story of D–O cycles includes a moment of intellectual discomfort that is worth lingering on. In the early years of ice-core analysis, the idea of repeated, large, abrupt temperature shifts ran against expectations shaped by slower geological thinking. Dansgaard and Johnsen (1993) presented a long stable-isotope record arguing for pervasive glacial instability, with the Holocene appearing exceptionally stable by comparison. To accept that picture required trusting that the ice was not lying: that the layers were correctly dated, that the signal was not an artifact of local snowfall changes, and that the abrupt transitions were not simply storms or seasonal effects masquerading as climate events.

The way that discomfort was resolved was not by rhetoric but by convergence. Multiple Greenland cores showed broadly similar patterns. Marine sediments showed correlated changes in the North Atlantic. Chemical tracers and dust indicators shifted alongside the isotopes. The same story was written in different inks. When independent archives agree on abruptness, the burden of proof shifts: the system really did change quickly, and the job becomes to understand why.

One of the most counterintuitive aspects of this story is that an abrupt event in Greenland can have a delayed partner in Antarctica. On first encounter, this feels like it should be impossible. How can opposite ends of the planet respond to the same event, sometimes with different timing and even opposite sign?

The bipolar seesaw offers a physically grounded answer. The deep ocean is a heat reservoir, and the overturning circulation is a conveyor of that stored heat between hemispheres. If the AMOC weakens, less heat is carried northward. The North Atlantic cools quickly. The heat that would have gone north stays in the Southern Hemisphere and the global ocean, which can allow Antarctica and the Southern Ocean to warm gradually. Stocker and Johnsen (2003) showed how a minimal thermodynamic model could reproduce aspects of this interhemispheric phasing with characteristic timescales on the order of a thousand years. The point is not that a toy model captures every detail, but that it makes the seesaw intuitive: a change in ocean heat transport does not simply cool the planet; it redistributes heat, with lags set by the deep ocean's slow adjustment.

This interhemispheric behavior is another reason ocean circulation is such a plausible switch mechanism. The atmosphere can transmit signals globally in days to weeks, but it cannot store heat for centuries. The ocean can store and transport, and it can do so in ways that naturally produce lags and asymmetries.

The phrase "tipping point" is sometimes used carelessly, as though the climate has a single cliff edge beyond which everything instantly collapses. The D–O record suggests something subtler and, in a way, more troubling: a system can be *close* to a threshold for long periods, during which it becomes jumpy. Near a threshold, a system can flicker. It can flip into a warm state and later fall back, not because external forcing is oscillating neatly, but because internal variability and feedbacks have become strong enough to shove it around.

This matters for interpretation. If the AMOC has multiple regimes, then a gradual change in background conditions—say, increasing freshwater input or reduced surface cooling—does not necessarily produce a gradual weakening. It can produce long periods of modest change punctuated by abrupt reorganizations. Even if the external forcing is smooth, the response can be lumpy.

It also matters for humility. D–O cycles do not come with a single agreed-upon trigger. Different models can produce abrupt warmings through different pathways: shifts in sea ice, changes in freshwater forcing, oscillations in deep water formation location, or interactions with the atmosphere. The paleoclimate archives show that abruptness occurred; the exact chain of causality may vary from event to event. A climate system can have a characteristic style of behavior without repeating the same script word for word.

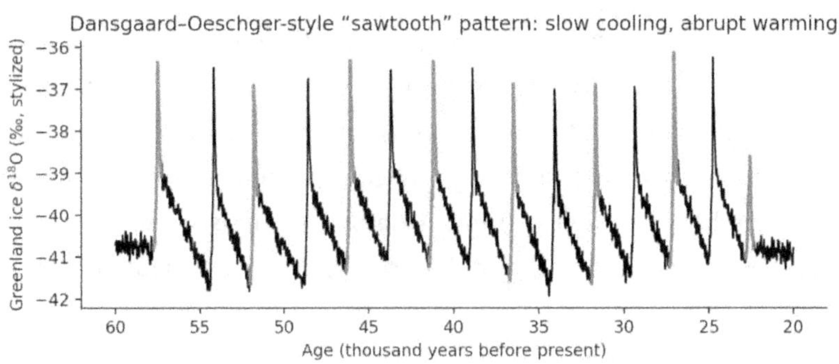

The chart is deliberately stylized, but it highlights the feature that makes D–O cycles so diagnostic: the near-verticalness of the warmings compared with the leisurely slide back toward cold. A system that responds to forcing like a simple thermostat does not typically produce that shape. A system that can occupy quasi-stable modes and flip between them can.

There is one more reason to take the flickering switch seriously: the present ocean already shows hints of how close the AMOC can ride to buoyancy thresholds. In winter, deep convection can be intense in some years and weak in others. Sea ice and freshwater export from the Arctic can vary substantially. Storm tracks shift. These variations do not produce D–O events today, but they demonstrate a principle: the ocean's ability to overturn is not guaranteed; it is negotiated each winter between surface cooling and surface freshening.

During glacial times that negotiation took place under very different terms. Ice sheets altered winds and storm tracks. Sea ice extended farther south. Freshwater sources were enormous. The North Atlantic was primed for abruptness. D–O cycles are the record of that priming: a climate system repeatedly nudged into one mode, then slipping back toward another, never settling into a smooth trajectory for long.

A switch that flickers is not broken in a trivial way. It is a switch whose internal feedbacks are strong enough that small disturbances can decide which state wins. In the Greenland ice, the winning state sometimes changed fast enough to feel instantaneous on the scale of a human life, leaving the impression, written in isotopes, of a world that could hold its breath for centuries and then exhale in a rush.

Chapter 6

Walking the Tightrope

Is the ocean's circulation a permanent fixture of our planet, or a fragile equilibrium balancing on the edge of mathematical probability? By reducing the complex Atlantic to its fundamental physics, we discover a startling duality where the current exists in two contradictory states-and the path between them is a treacherous one-way street.

6.1 A Tale of Two Boxes

Imagine trying to explain the Atlantic Ocean with something you could sketch on the back of a napkin at a crowded dinner party. Not the coastline or the currents you can see from space, but the hidden engine: the slow, muscular exchange in which warm surface waters drift northward, cool, sink, and return south at depth. The real Atlantic has eddies, storms, boundary currents, overflows, mixing over ridges, and water masses with names that sound like characters in a novel. Yet in 1961 Henry Stommel offered a model that was almost offensively simple: two boxes of water, connected by a pipe. From that toy came one of the most unsettling lessons in climate science: a circulation can have more than one stable way of operating under the same external conditions.

The two boxes stand in for two broad regions. One is low-latitude: the subtropical or equatorial Atlantic, where sunlight is strong and evaporation tends to exceed rainfall. The other is high-latitude: the subpolar North Atlantic, where winter cold bites, sea ice can form, and rain and snow can freshen the surface. The boxes are not meant to be literal containers. They are a way of focusing attention on what matters for the large-scale overturning: density, and the two main ingredients that set density at the sea surface, temperature and salinity.

Water, unlike air, can change its density substantially without changing its pressure very much, simply by being warmed or cooled, freshened or made saltier. Warmer water is lighter than colder water. Fresher water is lighter than saltier water. Those two facts are easy to remember, but they set up a competition that is not always intuitive. In the high latitudes, cooling tries to make surface water heavy, encouraging it to sink. Freshwater input from rain, rivers, and melting ice tries to make surface water light, resisting sinking. The overturning circulation lives in the tension between these tendencies: heat loss in the north pulls the system toward sinking and renewal, while freshwater can put a lid on the ocean, insulating it from the depths and slowing the engine.

Stommel's move was to turn that qualitative story into a minimal dynamical system. Give each box a temperature and a salinity. Connect them so that water can flow from one to the other and back. Decide that the flow rate depends on the density difference between the boxes: if the polar box is denser, flow goes one way; if the tropical box is denser, it goes the other. Then impose simple external influences: the low-latitude box is nudged toward being warm (by net heating) and salty (by evaporation), while the high-latitude box is nudged toward being cold (by net cooling) and fresh (by precipitation and ice melt). The details can be made more elaborate, but the spine of the model is that temperature and salinity both affect density, and the circulation itself transports heat and salt in ways that can reinforce or weaken the original density contrast.

It is tempting to think of this as an accountant's exercise: add up heat and salt, calculate density, deduce flow. The surprise is that the accounting can balance in more than one way. Stommel discovered that, for a range of external freshwater forcing, there can be two stable solutions: one with vigorous overturning, and one with weak or even reversed overturning. Both satisfy the same external constraints. Which one the system occupies depends on its history, like a mechanical switch that can sit in either of two positions.

To see why, it helps to name the two opposing influences in plain language. There is a *thermal drive*: the push created when high latitudes lose heat to the atmosphere and become denser than the tropics. And there is a *haline brake*: the tendency of freshwater input in the north to reduce salinity, lighten the surface, and prevent sinking. The circulation itself participates in this tug-of-war. When the overturning is strong, it carries warm water northward at the surface. That warmth seems, at first glance, as if it ought to *reduce* sinking by making the northern surface lighter. But it also carries salt northward, and salt is heavy. Which effect wins depends on the balance between temperature and salinity in setting density and on how the atmosphere responds.

One of Stommel's key insights was that salt transport creates a feedback that can be self-reinforcing. If surface flow from the tropics to the north is strong, the north receives more salt. Saltier water is denser, so it is more willing to sink, which supports strong flow. Strong flow brings more salt, and so on. This is the salt-advection feedback in its simplest form, and it is the reason the model can have two stable regimes. In the strong regime, the circulation keeps the north salty enough to sink despite freshwater input. In the weak regime, reduced northward salt transport allows the north to freshen, which further suppresses sinking and keeps the circulation weak.

An analogy can make the logic feel less abstract. Picture a public fountain with a pump that depends on water pressure. If the pump is running, it keeps pressure high; if pressure is high, the pump runs well. Once the pump slows, pressure drops; once pressure drops, the pump slows further. Now add a constant leak that drains the system. If the pump is strong enough, it can overcome the leak and maintain flow. But if a disturbance slows it past a certain point, the leak wins and the system settles into a low-flow state. Crucially, restoring conditions to what they were before the disturbance may not restart the pump unless you push it harder than you did on the way down. In Stommel's boxes, the leak is freshwater input; the pump is salt transport that helps maintain density in the north.

The historical context matters, because it explains why the simplicity was not a weakness but a strategy. Stommel was working at a time when the ocean was being mapped with a new seriousness. The Second World War and the early Cold War had poured resources into oceanography for navigation and submarine warfare. Instruments improved. Large-scale expeditions and long time series began to appear. Yet a comprehensive computational model of the global ocean was still out of reach. Even the idea that the ocean could have multiple stable states was not a standard part of geophysical thinking. The default intuition was that for a given forcing there would be a unique, predictable circulation. Stommel's two-box model was a deliberate provocation: a demonstration that the ocean's governing physics is nonlinear enough to allow *choice*.

Nonlinearity enters in a quiet way. The flow depends on density difference; density depends on temperature and salinity; temperature and salinity evolve under forcing and transport. Put those relationships together and you no longer have a simple proportional response. You have a system that can amplify small differences, or suppress them, depending on where it is in parameter space. In the language of dynamical systems, you can have multiple equilibria. In ordinary language, you can have two different "normal" Atlantics.

A concrete example helps anchor the abstraction in something people recognize: the contrast between a maritime winter and a continental one. Western Europe, at the latitude of Labrador, is famously milder than you might expect. Part of that is atmospheric circulation, and part is the ocean's delivery of heat. The Atlantic overturning circulation is often described, somewhat loosely, as a conveyor that carries warm water north and releases heat to the atmosphere. If that delivery weakens, regional climate patterns can shift. The two-box model does not predict exactly how Paris or Oslo would feel under a different overturning state, because it has no winds, no storms, no geography. What it does is establish a more basic point: there is no guarantee that the ocean's heat delivery changes smoothly as the climate changes. A gradual increase in freshwater input to the North Atlantic can, in principle, tip the circulation from a vigorous state to a weak one, altering the ocean's role in regional climate even if the external forcing continues to change slowly.

It is easy to misinterpret this as a claim that "freshwater turns the circulation off." The truth is subtler, and Stommel's model forces that subtlety into view. Freshwater does not push a button. It changes salinity and therefore density. Whether that change matters depends on how close the system is to the balance point where thermal drive and haline brake nearly cancel. In a strongly driven system, freshwater perturbations may be absorbed without drama. In a delicately balanced system, the same perturbation can be decisive. The model's power is not that it predicts a shutdown on a particular date, but that it gives a mechanism for *threshold behavior*.

The model also clarifies a common confusion about temperature versus heat transport. When the circulation is strong, it imports heat to high latitudes. That tends to warm the northern surface, making it lighter and opposing sinking. Does that not undermine the circulation? It can, but in Stommel's framework the atmosphere is continually cooling the northern box. The north is not warm *despite* heat loss; it is cool *because of* heat loss. The circulation's job is not to make the north cold; it is to supply heat that can be lost. In a strong overturning state, the north can lose more heat because it is continually resupplied. The thermal drive is not the *temperature* difference itself; it is the persistent tendency of the north to be cooled by the atmosphere relative to the south. That cooling creates dense water, but only if freshwater does not make the surface too buoyant to sink. The thermal drive and the haline brake act on the same lever: density.

Stommel's boxes also invite a more careful distinction between *salinity* and *freshwater*. Freshwater forcing is often described as if it simply dilutes the ocean, but in the real Atlantic it is an accounting of many processes: precipitation minus evaporation, river runoff, ice-sheet melt, sea-ice formation and melt, and the import or export of freshwater by ocean currents through gateways. The two-box model compresses all of this into a single parameter: a tendency for the polar box to be fresher than the equatorial box. That compression is exactly what makes the model useful. It tells you that, no matter how freshwater arrives, the ocean's response depends on whether the circulation can continue to deliver salt northward fast enough to maintain sinking.

A minimalist schematic captures the idea without drowning it in details. The warm, salty box in low latitudes tries to feed the cold, fresh box in high latitudes. A red arrow marks the thermal drive that encourages northward flow and sinking; a blue arrow marks the freshwater resistance that opposes it. Between them is the flow itself, drawn as a loop: surface transport one way, deep return the other. The point is not the geometry but the competition.

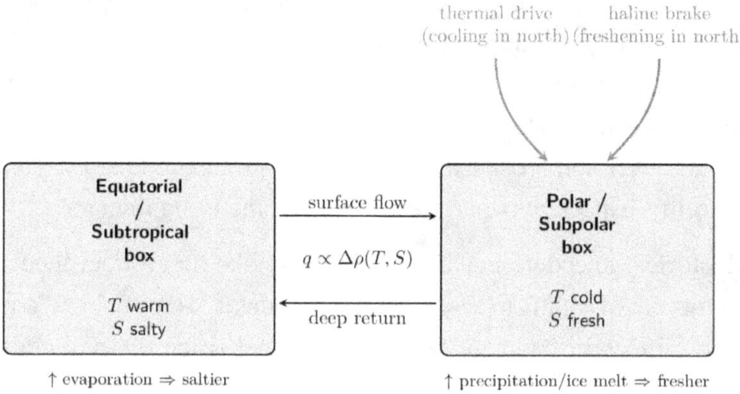

The equation at the bottom, $q \propto \Delta\rho(T, S)$, is the whole philosophy in one line: the flow rate q is proportional to the density difference between the boxes, and density depends on temperature T and salinity S. The rest is interpretation.

If you want a touch more mathematical intuition without heavy machinery, consider how density is often approximated:

$$\rho \approx \rho_0\big(1 - \alpha(T - T_0) + \beta(S - S_0)\big),$$

where α measures how strongly density decreases with warming and β measures how strongly density increases with salinity. The exact numbers depend on pressure and temperature range, but the form is robust: warming makes water lighter, salt makes it heavier. In the two-box model, the circulation "looks at" the difference $\Delta\rho$ between north and south. If the north is much colder than the south, $\Delta\rho$ tends to favor sinking in the north. If the north is much fresher, $\Delta\rho$ can shrink or even reverse.

The feedback enters because T and S are not fixed. They are continually pushed around by air—sea exchange and by the circulation itself. If the circulation weakens, less heat and salt are carried north.

Less heat transport might cool the north, which would strengthen density and oppose the weakening. Less salt transport would freshen the north, which would weaken density and amplify the weakening. Two feedbacks, opposite in sign, compete. Stommel showed that the salt-advection feedback can dominate strongly enough to create bistability, especially when freshwater forcing is significant.

A historical anecdote makes this feel less like an abstract theorem and more like a shift in scientific temperament. Stommel had a reputation for clarity and for taking conceptual risks. Oceanography, especially physical oceanography, can become enamored of detail: another current, another layer, another correction term. His two-box model was a refusal to be intimidated by complexity. It was also a wager that the public face of a scientific result is not its most complete equation set but its most revealing mechanism. In the decades that followed, far more detailed models were built, and the question "Can the Atlantic overturning have more than one stable state?" became a serious research program rather than a philosophical curiosity. But the two boxes remain on the desk, precisely because they can be held in the mind without a computer.

There is also a human reason the model resonates: it turns the ocean into something with a temperament. Not capricious, not magical, but capable of switching between modes once pushed hard enough. That possibility changes how one thinks about risk. If a system responds smoothly, then small errors in prediction lead to small errors in outcome. If a system has thresholds, then modest uncertainties can sit on top of large consequences. A two-box ocean is the simplest stage on which that drama can play out.

Yet the drama is not just about "collapse," a word that invites panic and oversimplification. The more unsettling message is that stability itself can be an achievement. In the strong overturning state, the circulation continually performs work: it moves heat, rearranges salt, ventilates the deep ocean, and maintains a particular pattern of density. The haline brake is always there, in the form of freshwater at high latitudes, and the thermal drive is always there, in the form of heat loss. The circulation persists because the feedbacks allow it to persist. In that light, the Atlantic overturning is less like a machine with a single setting and more like a balance that can be held, or lost.

Stommel's two boxes also sharpen an ethical intuition about climate change: not every consequence scales neatly with the size of the disturbance. Adding freshwater to the North Atlantic by accelerating ice melt is not merely "turning down" a knob. It is altering a control variable in a system that may have multiple resting places. The model does not say which resting place the real ocean will choose under a given future, but it does say that asking the question is scientifically legitimate. It gives a clean reason why "business as usual" thinking can fail in a nonlinear world.

On a cold winter night in the subpolar Atlantic, the surface ocean can cool rapidly, become heavy, and mix downward. This is not a metaphor; it is a physical event that sailors have felt as a change in sea state and that oceanographers have measured as deepening mixed layers and the renewal of deep waters. Now add a thin lens of fresh water at the surface, perhaps from persistent rainfall, perhaps from melting ice. The lens can be shallow enough to miss from a casual measurement and yet strong enough to keep the cold, dense water from connecting with the depths. A lid does not need to be thick to be effective. The two-box model is the distilled version of that night: cooling tries to make water sink, freshening tries to keep it afloat, and the outcome depends on whether the circulation keeps delivering salt to help the ocean overcome its own buoyancy.

That is why the story begins with two boxes. Not because the ocean is simple, but because the ocean's complexity can hide the few crucial levers. Temperature and salinity are those levers. Density is the hinge they turn. And once you see how a circulation can help maintain the very density contrast that drives it, you also see how it can fail to maintain it, settling into a different, quieter equilibrium in which the haline brake does not merely slow the engine but helps define a new normal.

6.2 The Hysteresis Trap

There is a particular kind of disappointment familiar to anyone who has tried to fix a stubborn household problem. The smoke alarm chirps because the battery is low. You replace the battery, but the chirp continues. You replace it again, you press the reset button, you wave a towel at the ceiling as if the alarm might be soothed by attention. Eventually you learn the rule: some alarms hold a little charge and need to be fully discharged, or a different contact is corroded, or the device itself is failing. The cause and the cure are not mirror images. The path back is not the path down.

Ocean circulation can behave in the same aggravating way. The overturning flow is not merely a response to today's pattern of heating and freshening; it can carry a memory of what happened before. In climate science the word for this is *hysteresis*, borrowed from physics. A system with hysteresis has more than one stable state under the same external conditions, and which state you end up in depends on how you got there. In the ocean, that means a circulation can weaken abruptly under increasing freshwater forcing, and then refuse to recover even when freshwater forcing is reduced to the earlier level. Restarting requires *more* than undoing; it requires pushing in the opposite direction past a second threshold.

The two-box picture already hints at why. A strong overturning state is self-maintaining partly because it transports salt northward, keeping high-latitude surface waters dense enough to sink. If the system is pushed into a weak state, that salt transport slackens, the high latitudes freshen, and the weak state becomes self-maintaining too. At that point, returning freshwater forcing to its previous value does not automatically bring back the lost northward salt transport. The system is caught in its own new consistency, like a well-worn groove.

The crucial step is to move from the idea of "two possible states" to the more unsettling idea of "two different thresholds." A single threshold is easy to imagine: push the system hard enough and it flips; relax the push and it flips back. Hysteresis replaces that symmetric picture with a loop. On the way down, the circulation can tolerate freshwater forcing until a critical point, and then collapse to a low-flow state. On the way back up, reducing freshwater forcing does not immediately restore the circulation; the system stays on the low-flow branch until a different, more demanding point is reached. Between those two points lies a range of forcing where *either* a strong or a weak circulation is possible, depending on history. That range is often called a *bistable regime*.

A helpful way to see the loop is to imagine you can control, like a dial, the net freshwater input to the North Atlantic: precipitation minus evaporation, plus runoff and ice melt. Turn the dial up slowly. The overturning weakens gradually at first, then suddenly drops. Now turn the dial down again. You might expect the circulation to trace its earlier path backward. Instead it stays weak and only recovers much later, after you have dialed freshwater down far below the value that triggered the collapse. The ocean has not "forgotten" how to overturn; it has reorganized itself so that the weak state now makes sense.

Stefan Rahmstorf's work in the mid-1990s gave this idea a sharper, more Atlantic-shaped form. In 1995 he explored how the Atlantic thermohaline circulation responds to changes in the hydrological cycle and found clear threshold behavior in models: moderate shifts in freshwater forcing can induce abrupt transitions between equilibria. The striking feature was not only the abruptness but the asymmetry: the conditions that produce shutdown are not the same as those required for recovery. That asymmetry is the hysteresis trap.

The phrase "trap" is not meant to be melodramatic; it is meant to be practical. A trap is a region of state space that is easy to fall into and hard to climb out of. The concern is not simply that the circulation could weaken, but that once weakened beyond a certain point it might resist recovery on human timescales even if the forcing is later reduced. This is why discussions of "reversibility" matter in climate risk: the costs of crossing a threshold are not necessarily paid only at the moment of crossing. They can show up later, as a stubborn refusal of the system to return.

A concrete example of hysteresis, outside the ocean, sits in many kitchens. A thermostat with a narrow hysteresis band prevents a heater from turning on and off every few seconds. The heater might switch on when the room falls to 19°C and switch off when it rises to 21°C. At 20°C the system could be either heating or not heating, depending on whether the temperature is rising or falling. That "dead band" is hysteresis, deliberately designed for stability. The ocean's hysteresis is not designed; it is an emergent property of feedbacks between circulation, salinity transport, and surface buoyancy.

The Atlantic's loop can be drawn as a simple curve: circulation strength on the vertical axis, freshwater forcing on the horizontal axis. The strong circulation occupies an upper branch. As freshwater forcing increases, the upper branch persists until it reaches a cliff edge, where it becomes unstable and the system falls to a lower branch. As freshwater forcing decreases again, the system remains on the lower branch until it reaches a different cliff edge, where the lower branch can no longer persist and the circulation jumps back up. Between the two cliff edges, both branches exist. In that bistable regime, a small nudge can matter, but so can the past.

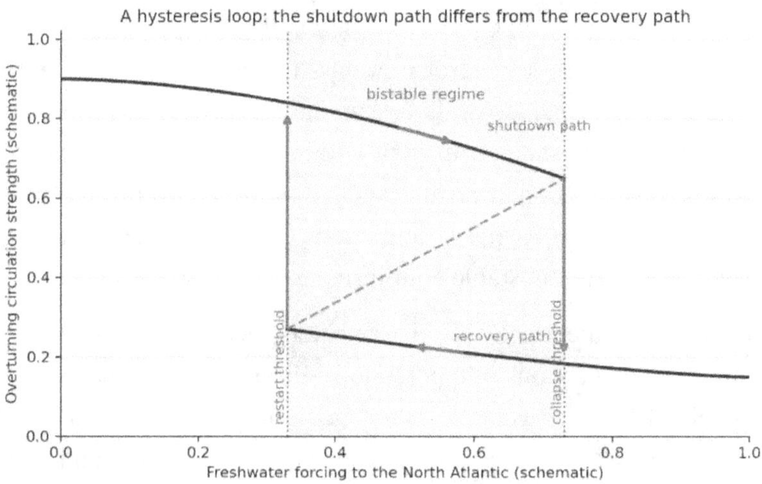

The figure is deliberately schematic, but the logic is real. A mild increase in freshwater forcing can be absorbed while the circulation stays on the upper branch, because the strong state actively maintains the salinity structure that supports sinking. Once pushed past the collapse threshold, the system drops to a low state in which weak salt transport allows freshening to persist. In the bistable regime, there is no single answer to the question "What is the circulation strength for this freshwater forcing?" The answer depends on whether the system approached from lower forcing or higher forcing.

That is the most important consequence of hysteresis for how we think about climate intervention. It changes the meaning of "restore." If a system is on the upper branch, keeping it there may require relatively modest action. If it has fallen to the lower branch, restoring the previous forcing may not be enough. The amount of effort required to bring the circulation back can be substantially larger than the effort required to keep it from falling in the first place.

It also changes the meaning of evidence. Suppose two decades show similar freshwater forcing in the North Atlantic, but the circulation behaves differently. In a purely single-equilibrium world that would demand a measurement error or a missing variable. In a hysteretic world it is not a paradox. Two "similar" climates can support different circulation states if they sit within the bistable window. The ocean can look calm from the outside while being poised near a boundary between basins of attraction: regions of state space that flow toward different stable equilibria.

This is where the emotional charge enters, because hysteresis turns gradual human actions into potentially abrupt natural consequences. Greenland melt, for example, is not an on/off switch; it is a changing freshwater flux. Yet the ocean's response to freshwater can be nonlinear. In a hysteretic system, you can watch a trend unfold slowly for years—warming air, thinning ice, increasing runoff—and still be surprised by a sudden change in circulation that arrives on the ocean's own timetable.

There is a well-known historical episode that captures the flavor of that surprise, even though it predates modern measurements: the 1960s "Great Salinity Anomaly." Oceanographers in the North Atlantic observed a large, persistent patch of unusually fresh water that propagated through the subpolar gyre over several years. The anomaly is not identical to an AMOC shutdown, and it had multiple contributing factors, but it became a vivid demonstration that salinity in the North Atlantic can reorganize on basin scales and on timescales short enough to matter for climate variability. It also shaped habits of mind in the community: freshening is not just a local dilution near a glacier; it can become a mobile, coherent feature that influences stratification and convection far downstream.

What hysteresis adds to that picture is an explanation of why freshening can have *after-effects*. If a fresh anomaly reduces deep mixing for a few winters, the water column structure changes. Reduced mixing can trap heat at depth or keep freshwater at the surface. In the real ocean, winds, eddies, sea ice, and boundary currents complicate the story, but the same logical core remains: the circulation affects the salinity distribution that controls sinking, and that coupling can store memory.

Modern assessments, such as the IPCC's Special Report on the Ocean and Cryosphere in a Changing Climate (2019), take this possibility seriously while also emphasizing uncertainty. The models do not all agree on how close the present-day Atlantic is to a threshold, and different models exhibit different degrees of hysteresis under freshwater perturbations. That model dependence is not a reason to dismiss the mechanism; it is a reason to be careful about what is robust. The robust part is the physics of buoyancy and feedback. The uncertain part is exactly where the real system sits relative to the bistable window, and how fast it might move under changing forcing.

The word "trap" also hints at a second complication: time. Hysteresis diagrams are often drawn as if the system always sits on an equilibrium curve. Real forcing changes in time, sometimes rapidly. When forcing changes faster than a system can adjust, tipping can be *rate-induced*: the system can cross from one basin of attraction to another even if the forcing never exceeds the equilibrium threshold. Work by Alkhayuon and Ashwin and colleagues (2019) made this point sharply in idealized models: the speed of change can matter as much as the magnitude. For ocean circulation, that means a rapid pulse of freshwater—a surge of meltwater, an extreme season of precipitation, a sequence of anomalous winters that suppress convection—could in principle push the system into a different state more efficiently than a slow, smooth change of the same total amount.

Rate effects are not an extra flourish; they match lived experience. People do not respond to the same news in the same way if it arrives as a gradual drift or as a sudden shock. Physical systems are similar. Slow changes allow adjustment, mixing, and compensating feedbacks. Fast changes can bypass those gentle stabilizers and push the system across a boundary before it has time to respond. In the ocean, the relevant adjustment times include the time it takes for salinity anomalies to be advected and mixed, and the time it takes for deep waters to be renewed. Those are not instantaneous. They are the ocean's internal clocks.

Then there is coupling: the Atlantic overturning does not sit alone. Greenland meltwater depends on atmospheric and oceanic heat transport; sea-ice extent depends on both ocean heat delivery and atmospheric circulation; rainfall patterns depend on the circulation of the atmosphere, which in turn responds to sea-surface temperatures influenced by the ocean. When subsystems are coupled, hysteresis can become more intricate than a single loop. Coupled tipping and cascading effects, increasingly explored in the literature (including work such as Klose and colleagues in 2024), can create one-way pathways that do not resemble a neat, symmetric diagram. In that broader picture, the "restart threshold" is not only about turning down freshwater forcing; it is also about the altered context in which the ocean would attempt to restart: different ice cover, different winds, different background stratification.

For a popular narrative it is tempting to turn hysteresis into a moral lesson: act early, because the return trip is harder. That lesson is not wrong, but it can sound preachy if it floats above the mechanism. The mechanism is more interesting and more human. A strong overturning state is, in a sense, a cooperative arrangement between ocean and atmosphere. The atmosphere cools the north; the ocean supplies heat and salt; dense water forms and sinks; the deep ocean is ventilated. A weak state is a different arrangement: the atmosphere can still cool, but the surface remains capped by freshwater; heat may accumulate below; salt transport slackens; deep renewal slows. Each state, once established, tends to reproduce the conditions that make it viable. That is why undoing the initial push does not necessarily undo the outcome.

Seen in that light, hysteresis is not a catastrophe in itself. It is a warning about the geometry of cause and effect. The world is full of systems that behave this way: ecosystems that flip from clear-water to turbid lakes, financial markets that can seize and then require extraordinary liquidity to restart, relationships that can fracture in a moment and take years to rebuild. The ocean is not moral, but it is not simplistic either. It can be resilient in one range and stubborn in another.

The most unsettling part of the hysteresis trap is how ordinary it can look from the inside. In the bistable regime, the strong circulation can appear robust right up to the threshold. The weak circulation, once established, can appear stable too. Neither state needs to announce itself with dramatic variability; stability can be the mask of proximity to a cliff. That is why the diagram matters: it trains the eye to expect that the "same" forcing can host different realities, and that the direction of travel changes the meaning of what you measure.

In the North Atlantic, winter after winter, the surface either connects with the depths or it does not. Salt either arrives fast enough to keep the surface heavy or it does not. A system with hysteresis turns those seasonal decisions into a longer memory. The ocean can keep a record not in words but in stratification: layers of light water stacked over heavier water, a quiet barrier that must be overcome before sinking can resume. Once that barrier is in place, restarting the overturning is less like turning a knob and more like breaking a seal.

6.3 Mapping the Cliff Edge

A tightrope walker rarely falls because gravity suddenly becomes stronger. The fall happens because the margin for correction narrows until the smallest wobble can no longer be rescued. For the Atlantic overturning circulation, the modern question is not whether the ocean obeys buoyancy physics—it always does—but where the margin has shrunk to something dangerously thin. *Mapping the cliff edge* means translating the elegant loop of a hysteresis diagram into the messy Atlantic of storms, overflows, sea ice, and meltwater, and then asking an uncomfortably practical question: how close is the real system to the point where the strong circulation state ceases to be available?

In the simplest models the cliff edge is literal: a point on a curve where the upper branch ends. Mathematicians call it a *bifurcation*, a change in the number or stability of equilibria. In the Stommel-style picture, as freshwater forcing grows, the strong overturning equilibrium persists until a turning point, beyond which it cannot exist. Past that point the system must jump to the weak state. The term *Stommel bifurcation* has come to stand for this loss of the "on" state in density-driven overturning: a threshold where feedbacks no longer support a vigorous circulation.

The Atlantic does not have a dial labelled "freshwater forcing," but it does have a freshwater budget, and it is changing. Evaporation in the subtropics, precipitation in the storm tracks, river runoff, Arctic export through gateways, sea-ice formation and melt, and Greenland ice loss all contribute. The difficulty is that the circulation reacts not only to how much freshwater arrives, but to where and when it arrives and to what else is happening at the same time. One can pour the same total amount of freshwater into two different places and get two different outcomes, because stratification—the layering of light water over heavy water—is local, and deep convection is local. Yet the consequences can be basin-wide, because the AMOC is a connected system: suppress sinking in one region and the entire overturning can adjust.

The temptation, in public discussion, is to hunt for a single number: *How many gigatons of ice melt will shut the AMOC down? How many degrees of warming? By what year?* A scientist's instinct is to reply, honestly, that the answer depends. But "depends" need not mean "nothing can be said." Mapping a cliff edge is not about predicting a date on a calendar; it is about identifying the geometry of risk. Where are the unstable boundaries? How wide is the safe corridor? How does that corridor shift as the climate changes?

A useful mental tool is the idea of a *basin of attraction*. Imagine releasing a marble on a gently contoured surface. Depending on where you let it go, it will roll into one valley or another. Each valley is a stable state. The ridge between valleys is the boundary between basins of attraction. A system can cross that boundary not only by changing the landscape (altering the climate forcing) but also by receiving a shove (variability, pulses, sequences of unusual winters). In equilibrium diagrams the ridge is represented by an unstable branch. In the real ocean it is a complicated, high-dimensional boundary in a space defined by temperature, salinity, currents, and stratification patterns. The cliff edge is where the ridge moves so close to the present state that ordinary fluctuations can push the system across.

A concrete example makes this less abstract. Consider deep water formation in the Labrador and Greenland-Iceland-Norwegian seas. In winter, cold air outbreaks can strip heat from the sea surface and create dense water that mixes downward, sometimes to great depth. Now add a persistent freshening at the surface. Fresh water is buoyant; it forms a cap that resists mixing. If the cap is strong enough, even intense cooling can fail to punch through to depth. That does not necessarily shut down the AMOC instantly, because the overturning involves several pathways and the ocean can compensate by shifting where dense water forms. But it can weaken the system's resilience. A few winters in which convection is suppressed can allow heat and freshwater to accumulate in the upper ocean, making it even harder for convection to resume. That is basin geometry in action: the state is nudged toward the ridge, and recovery becomes harder.

Historically, oceanographers learned to respect such reorganizations through episodes that looked, at first, like curiosities. One is the so-called Great Salinity Anomaly of the late 1960s and 1970s, when a large body of unusually fresh water propagated around the subpolar North Atlantic. Another is the dramatic changes in deep convection in the Labrador Sea across the late twentieth century, with periods of intense mixing followed by decades when deep convection weakened. These events are not direct evidence of a looming Stommel bifurcation, but they are reminders that salinity anomalies can persist and spread, and that convection can switch between regimes. The ocean has modes.

The modern challenge is to connect that mode-switching intuition to a measurable indicator of proximity to a tipping point. A bifurcation, in a mathematical model, can be located precisely. In the real ocean, we have incomplete observations, changing measurement systems, and a climate background that is itself drifting. So scientists have developed two broad families of warning approaches.

One family is statistical. Many systems slow down as they approach a critical transition: after a perturbation they take longer to recover. In time series this can appear as rising autocorrelation (today looks more like yesterday) and rising variance (fluctuations grow). Niklas Boers (2021) applied this logic to several observational proxies for AMOC variability derived from sea-surface temperature and salinity patterns and reported early-warning signals consistent with declining stability over the last century. The interpretation is straightforward:

if the AMOC is approaching a tipping point, it may become more sluggish in returning to its mean state.

But the ocean is not a laboratory pendulum, and statistics can be fooled. A critique published in *Nature Climate Change* in 2023 argued that changes in observational coverage and sampling can create spurious trends in variance and autocorrelation, producing apparent early-warning signals even when the underlying dynamics do not contain them. This is not a technical quibble; it is a warning about a deeper pitfall. When the observing system changes—satellites come online, ship tracks shift, profiling floats are deployed—the character of the data changes too. An apparent increase in variability can reflect the world becoming more variable, or it can reflect the world being better observed. Untangling those is hard, and the stakes are high.

The second family of warning approaches is physics-based: instead of looking for generic statistical fingerprints of critical transitions, look for a specific quantity that the theory says controls stability. For the AMOC, one of the most discussed candidates is the freshwater transport by the overturning circulation at the southern boundary of the Atlantic, often near 34°S. The intuition is tied to the salt-advection feedback. If the overturning imports salt into the Atlantic basin (equivalently, exports freshwater), that tends to support the strong state. If it does the opposite—imports freshwater—the feedback can flip sign, making the strong state less stable. In 2024, René van Westen and colleagues explored a stability indicator built around this freshwater transport in simulations with a comprehensive climate model under freshwater forcing. Their work helped crystallize a key point: stability may be more directly tied to the sign and strength of freshwater transport than to any single local measurement in the North Atlantic.

This is conceptually satisfying because it feels like the ocean doing bookkeeping at basin scale. The Atlantic is not just a place where water sinks; it is a basin with boundaries. Freshwater can enter from the atmosphere and continents, but it can also enter or leave through ocean gateways, carried by currents. The overturning itself is part of that exchange. If the overturning's freshwater transport changes sign, the circulation can cease to be self-sustaining in the Stommel sense.

Even here, caution is required. Measuring freshwater transport at 34°S with the needed accuracy is difficult, because it involves not only mean flow but eddies and the division between overturning and gyre components. And models have biases: they may represent rainfall patterns, winds, mixing, and boundary currents imperfectly. The very existence of a multiple-equilibria regime can be sensitive to those biases. Work by Dijkstra and van Westen (2023) emphasized that biases in surface freshwater fluxes can shift the model's bistable regime into parameter ranges that may be unrealistic. That does not mean bistability is impossible in the real world; it means that a model can possess the mechanism in principle but place the cliff edge in the wrong place. Mapping the cliff edge, then, becomes a two-part task: understand the mechanism, and then understand the model errors well enough to know whether the mechanism is relevant under realistic forcing.

The phrase "how close are we?" is therefore not one question but several. How close is the Atlantic's present freshwater budget to the modelled bifurcation threshold? How close is the present state, in the space of temperatures and salinities, to the boundary between basins of attraction? How close is the *trajectory* of the system to a rate-induced threshold, where a rapid change could tip it even if the equilibrium threshold is not crossed?

Rate matters because the ocean has internal inertia. Deep waters form, spread, and mix on timescales of decades to centuries. Surface forcing can change on timescales of years to decades. If forcing changes quickly compared to the system's adjustment time, the ocean can be pushed out of the basin of attraction of the strong state. The work of Alkhayuon, Ashwin and collaborators (2019) highlighted how time-dependent forcing can induce tipping without exceeding the static bifurcation value in a global box model. This adds a sobering nuance to policy-relevant thinking. A particular end-state of greenhouse gas concentrations or freshwater forcing is not the only thing that matters; the pathway and speed of change matter too.

Here is an everyday analogy that is close enough to be useful. Think of a bicycle moving slowly. At moderate speed it is stable; small wobbles are corrected almost automatically. As you slow down, the bike becomes harder to balance. There is no single speed at which balancing becomes impossible for all riders, because skill and conditions matter. But there is a clear change in geometry: the basin of attraction for upright riding narrows. A gust of wind that would be harmless at 15 miles per hour can topple you at 2. The AMOC, under increasing freshwater and warming, may be like the bicycle slowing: the same variability that used to be absorbed can become destabilizing.

This brings the story to the present day, and to the question of what scientists mean when they talk about an AMOC "tipping point." In the most careful usage, a tipping point corresponds to a qualitative change in system behaviour that is difficult to reverse on the timescale of interest. In Stommel's vocabulary, it is the loss of the strong equilibrium or a transition between basins of attraction. In public conversation, the phrase sometimes suggests a sudden stop of all currents, which is misleading. The Gulf Stream, a wind- and density-driven boundary current system, will not simply halt in a day. The AMOC is an overturning component, not the entire surface circulation. Yet it can weaken sharply in models under strong freshwater perturbations, and such weakening can have large climate impacts even if "collapse" is not absolute.

There is also a psychological trap in the way humans picture cliff edges. We imagine walking calmly until the ground ends. But in many dynamical systems the cliff edge is more like a steepening slope: subtle signs can appear long before the drop, and the moment of tipping is often preceded by changes that are easy to overlook unless you know to look for them. In the ocean those signs could include shifts in salinity patterns, changes in deep convection frequency, changes in freshwater transport, and altered relationships between wind forcing and overturning. None of these is a single smoking gun. Together they form a pattern, but the pattern is blurred by natural variability.

A small historical anecdote from the culture of oceanography captures how hard it is to map the cliff edge with limited data. For much of the twentieth century, scientists inferred large-scale circulation from sparse hydrographic sections: a ship would cross the Atlantic once, measure temperature and salinity profiles, and then return years later. Enormous inferences were made from those snapshots, often with admirable ingenuity and sometimes with overconfidence. The modern era of continuous monitoring—arrays like RAPID at 26.5°N, satellite altimetry, Argo floats—has transformed the field. Yet even now, a few decades of direct observations are short compared with the ocean's memory. The cliff edge could be close or far, and the system could still look deceptively familiar, because "familiar" is defined by the last few human generations of measurements.

There is one more reason mapping is difficult: the cliff edge itself can move. The bifurcation point in a model is not a permanent feature of nature; it depends on background conditions such as temperature structure, wind patterns, mixing rates, and the geometry of ice and coastlines. As the planet warms, those background conditions evolve.

Warming can increase stratification by making surface waters lighter. Changes in winds can alter gyre transports and mixing. Greenland meltwater can freshen key regions. All of these can shift the location of the bifurcation, potentially bringing it closer without requiring a dramatic additional perturbation. In that sense, the cliff edge is not a fixed precipice but a shifting one.

If this sounds unsettling, it is worth remembering that uncertainty cuts both ways. Some models may overstate sensitivity, placing the cliff edge too close because of biases in freshwater fluxes or mixing. Others may understate sensitivity if they fail to represent key processes or underestimate feedbacks. The ocean's true stability is not a number to be announced; it is a hypothesis to be tested from multiple angles. The healthiest posture is neither complacency nor certainty, but a kind of scientific vigilance: insist on indicators tied to physics, insist on robust statistical methods that account for changing observations, and insist on understanding model biases rather than averaging them away.

The tightrope image becomes most useful when it is made operational. A tightrope walker watches not the abyss but the rope: its tension, its sway, the subtle changes that precede a big movement. For the AMOC, the rope is the coupled pattern of buoyancy forcing and salt transport that sustains deep water formation. When freshwater input increases, the haline brake presses down. When the overturning weakens, salt import slackens, and the brake can press harder. The cliff edge is where that mutual reinforcement wins.

In the end, mapping the cliff edge is an attempt to answer a simple human question with honest physics: *If the Atlantic overturning is pushed, will it come back the way it was?* Hysteresis says "not necessarily." Bifurcation theory says "sometimes no." Basin geometry says "the shove matters." Observations say "the system varies, and the measurements are imperfect." Together they paint a picture that is neither a prophecy nor a shrug. It is a warning about the shape of possibility: a world in which gradual change can narrow resilience until a once-rare outcome becomes an easy one, and where the absence of drama is not proof of safety but sometimes only the calm of a system still on its upper branch.

Chapter 7

Taking the Ocean's Pulse

For decades, the deep Atlantic remained a black box, illuminated only by the rare flashlight of a passing research vessel. Today, a vast network of tethered sensors and autonomous robots has transformed the ocean into a transparent medium, revealing a circulation system that is far more volatile than theory predicted. We examine the machinery of this observational revolution and the vital data it streams to our climate models.

7.1 The Gatekeepers at 26 North

For most of the twentieth century, the Atlantic Meridional Overturning Circulation (AMOC) was something oceanographers inferred rather than observed. Ships crossed the Atlantic, stopped at stations, and lowered instruments to measure temperature and salinity. Those numbers were then turned into density, and density into pressure gradients, and pressure gradients into currents. It was ingenious, and it worked well enough to sketch the ocean's broad architecture. But it was also like estimating the traffic on a motorway by taking one aerial photograph every few years and hoping the rush hour never changes.

That hope quietly shaped the science. The AMOC was often described as a slow, steady conveyor: a northward flow of warm surface water balanced by a southward return flow at depth. If it weakened, the story went, it would do so gradually, over decades. Yet that confidence rested on a thin observational record. A ship section might tell you the flow on that particular week; it could not tell you whether the ocean had been doing something entirely different the week before.

At 26.5°N, roughly the latitude of the Canary Islands and the Florida Keys, the Atlantic offers a rare gift to observers: a place where geometry and politics cooperate. The basin is wide but bounded; major currents squeeze through predictable corridors; and there are coastlines and islands on both sides that can support logistics. In the early 2000s, a UK—US collaboration built an observing system there with an audaciously simple aim: measure the AMOC continuously, not as a one-off expedition trophy but as a living time series.

The result was the RAPID—MOCHA—WBTS array, commonly shortened to RAPID. The acronym sounds brisk and modern, but its achievement is closer to stubborn persistence: keeping a picket fence of instruments alive in a corrosive, high-pressure, storm-driven ocean, year after year, long enough for the ocean to reveal its habits.

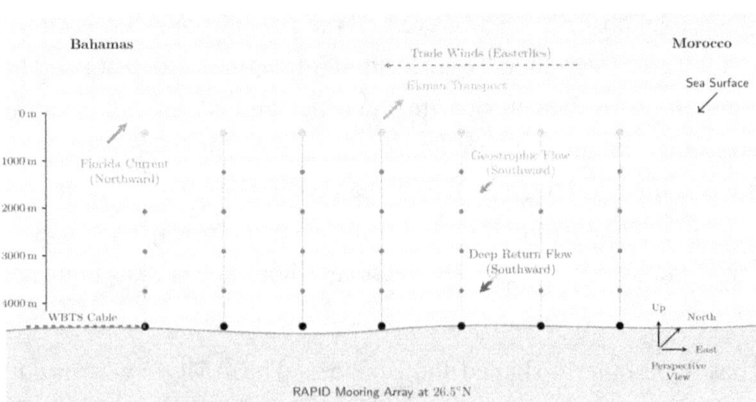

A useful way to picture RAPID is not as a single instrument but as a budget, balanced every day. The AMOC transport at 26.5°N is estimated by combining three ingredients.

First is the Gulf Stream, the fast boundary current that shoots north through the Florida Straits. Conveniently for science, the Straits are narrow and heavily trafficked by communications infrastructure. A submarine cable laid across the seafloor can be used as an electromagnetic flow meter: moving seawater in Earth's magnetic field induces a tiny voltage, and that voltage is related to the volume transport through the channel. This is the WBTS component (Western Boundary Time Series), and it anchors the AMOC estimate to one of the strongest, best-constrained currents in the world ocean.

Second is the wind. When trade winds blow across the Atlantic, they push the surface ocean, and because of Earth's rotation the net transport is angled and organized. Oceanographers call the resulting depth-integrated flow in the top few tens of metres *Ekman transport*. It is not a current you can point to with your finger like the Gulf Stream; it is a bookkeeping term that turns wind stress into a slab of water moving sideways. At 26.5°N, this wind-driven component can add or subtract several million cubic metres per second from the northward flow. That number is not a typo: ocean circulation is often measured in Sverdrups, where $1\,\mathrm{Sv} = 10^6\,\mathrm{m}^3\,\mathrm{s}^{-1}$. A large river might carry 0.01 Sv. The Atlantic carries tens of Sv.

Third is everything else in between Florida and Africa: the broad interior flow that is not captured by the Straits measurement. Here the moorings do the heavy lifting. Tall lines are anchored to the seabed and held upright by buoyancy. Along each line sit instruments that measure temperature and salinity (from which density is calculated) and pressure at depth. With enough vertical information on both sides of the basin, the density field gives the slope of pressure surfaces; the slope gives the *geostrophic* shear (the balance between pressure gradients and the Coriolis effect); and that shear can be integrated to obtain transport. It is a chain of reasoning, but it is a tested one: it is the same physics that lets weather maps translate pressure patterns into winds.

RAPID was designed so that these pieces sum to the total, day after day, with a careful correction to ensure that mass is conserved. If the northward flow in the upper ocean strengthens, there must be a compensating southward flow somewhere else, and the array is built to see that compensation rather than assume it.

What made this worth the trouble was the question that hovered over climate science at the time: how stable is the overturning circulation? Models had long suggested that the AMOC could weaken under greenhouse warming as the North Atlantic surface freshened and warmed, making it harder for surface waters to become dense enough to sink. But the models also differed wildly in how fast and how smoothly such a weakening would occur. Without continuous observations, it was easy for both optimists and pessimists to project their expectations onto sparse data.

RAPID's early years brought a surprise that was less cinematic than a sudden shutdown but, for scientists, almost as unsettling: the AMOC *wobbles*. In 2007, Stuart Cunningham and colleagues, analysing the first year of continuous measurements, showed that the transport could swing strongly on timescales of days to weeks. That finding did not mean the ocean was becoming unstable in the tipping-point sense; it meant the earlier observational strategy had been systematically underestimating the ocean's variability. A single ship section, taken in a particular season, could easily land on a temporary high or low and be mistaken for a longer-term state.

To appreciate why that mattered, imagine trying to detect a slow decline in a person's health by measuring their pulse once a year, always on a different day, sometimes after a sprint, sometimes after sleep. The pulse is real, but the noise is large. RAPID did not eliminate the noise; it revealed it, which is a different kind of progress. Once variability is acknowledged, the task becomes harder but also more honest: trends must be detected *through* the weather of the ocean, not by pretending it is calm.

The public first heard of RAPID's variability through a dramatic episode in 2009—2010, when analyses showed a substantial temporary reduction in the overturning strength. Headlines flirted with catastrophe, as headlines do. Within the scientific community, the episode played a more constructive role. It became a test case for the question RAPID was built to answer: is a sudden-looking change a harbinger of long-term decline, or simply the ocean's normal range of motion? With a continuous record, the event could be placed in context: it was large, but it was also reversible on the timescales observed, bound up with wind patterns and boundary current shifts rather than a one-way slide into a new regime.

Those years also forced clarity about what "measuring the AMOC" really means. The overturning circulation is often drawn as a single loop, but the measurement at 26.5°N is an accounting of northward and southward transports across that latitude, separated by depth (and sometimes by density classes). When the AMOC varies, it can do so by changes in the Gulf Stream transport, changes in wind-driven Ekman flow, or changes in the interior geostrophic flow. RAPID made it possible to see those contributions separately. That decomposition matters because the climate implications are not identical. A wind-driven fluctuation can be quick and energetic, yet carry a different signal for long-term heat storage than a persistent change in deep return flow.

If the physics is subtle, the engineering is blunt. The North Atlantic is not gentle on equipment. Moorings sit in salty water under immense pressure, abraded by currents, and threatened by fishing gear. Instruments foul, batteries die, lines chafe. Even finding a mooring again after a year can be non-trivial. Retrieval is often a small drama at sea: a ship returns to a set of coordinates, sends an acoustic signal downward, and waits for an answer from an underwater release mechanism. If the release works, the mooring detaches from its anchor and the buoyant line rises, sometimes taking an hour or more. The first sign of success can be a bright float bobbing in the waves, a small, almost comic object for something tasked with monitoring a planetary-scale current.

There is a particular kind of tension in such work that rarely reaches the polished figures of journal articles. Servicing cruises are expensive and scarce. A storm can steal days. A failed recovery can mean a hole in the time series that no clever analysis can fully repair. When the reward is "data continuity," it can be difficult to explain to outsiders why a missing month matters. Yet for detecting changes against a variable background, continuity is the whole point. A gap is not just an inconvenience; it is an invitation for the ocean to do something interesting while no one is looking.

The array also taught a lesson about the difference between *monitoring* and *understanding*. Monitoring means you can say what the AMOC did last week. Understanding means you can explain why it did it, and what it will do under conditions that have not yet occurred. RAPID pushed scientists from the first toward the second by giving them a reality check. Climate models could no longer be judged only by whether they produced a plausible average overturning strength; they could be challenged on whether they produced the observed amplitude and timing of variability at weekly, seasonal, and interannual scales.

Seasonality, in particular, became less of an afterthought. Earlier pictures of the AMOC often implied a slow, almost geological flow.

RAPID showed a strong seasonal cycle superimposed on shorter flickers. Thomas Kanzow, Cunningham, and colleagues analysed several years of data and quantified these seasonal swings, providing a concrete target for models: not just a mean, but a calendar of strengthening and weakening. That matters because the ocean's job in climate is inseparable from timing. A surge of northward heat transport in late autumn has different consequences for winter storm tracks than the same surge in midsummer.

A deeper, more human consequence of RAPID's record is the way it changed scientific temperament. When observations are sparse, theory tends to fill the gaps with confidence. When observations are dense, confidence becomes conditional. Researchers learned to speak in probabilities, error bars, and ensemble ranges not because they became timid, but because the ocean had earned their respect. The AMOC could not be treated as a single number that slides smoothly up and down; it behaved more like a system with moods, driven by winds, boundary currents, and internal adjustments that can amplify or cancel one another.

None of this diminishes the climate concern that motivated RAPID. If anything, it sharpens it. A variable system can hide a trend for a long time, and then make it seem suddenly obvious. It can also generate alarming excursions that later relax, tempting people to over-interpret the latest dip or spike. Continuous monitoring does not provide comfort; it provides discipline. It forces questions to be asked in the language of time scales: how much of today's weakness is weather, how much is season, how much is a persistent shift? And it forces a parallel question in the language of mechanisms: is the change being driven from above by winds and buoyancy fluxes, or from the sides by boundary current dynamics, or from below by adjustments in deep water masses?

A final irony is that the more clearly RAPID drew the line at 26.5°N, the more obvious it became that a line is not the whole story. The AMOC is not a single valve located at one latitude. It is a connected set of transformations: warm water moving north, losing heat to the atmosphere, being made denser by cooling and salinity changes, sinking and returning south at depth. RAPID's genius was not that it captured every detail, but that it pinned down one crucial cross-section well enough to reveal both the power and the restlessness of the circulation.

Standing on the deck of a research vessel at night, watching a recovered instrument drip and clatter as it comes aboard, it is hard to avoid a sense of scale mismatch. The sensor fits in a person's hands; the current it helps measure carries more water than all the rivers on Earth combined. Yet climate science often advances exactly through such mismatches: small, carefully calibrated devices turning the planet's largest flows into numbers that can be argued about, improved, and, eventually, trusted. At 26.5°N, the Atlantic was no longer a rumour reconstructed from occasional crossings. It became something closer to a monitored heartbeat: irregular, energetic, and impossible to dismiss as merely background.

7.2 The Robotic Flotilla

A line of instruments across one latitude can feel like a decisive experiment: a gate, a tollbooth, a place where the ocean must show its papers. But the North Atlantic is not a single corridor. North of the subtropics, the sea floor rises into ridges and sills, the coastline fractures into shelves and fjords, and the currents split, merge, and wrap around basins like weather systems. The water that eventually returns south in the deep limb of the AMOC does not sink at one neat point; it is transformed through a messy sequence of cooling, mixing, and recirculation. If 26.5°N is a sensible place to keep score, the subpolar Atlantic is where the game is played.

That is why the observing strategy had to change as scientists moved their attention northward. In the subtropics, the Atlantic is wide and comparatively simple: a strong western boundary current, a broad interior, and a deep return flow. In the subpolar gyre—the counterclockwise circulation spanning Greenland, Iceland, the Labrador Sea, and the waters south of them—the path of overturning is braided. Warm, salty Atlantic water enters along multiple branches, meets fresher Arctic outflows, and is stirred by intense storms and sharp fronts. Deep waters form in several places and are then steered by the topography of the Mid-Atlantic Ridge and the Greenland–Scotland Ridge. Measuring overturning here is less like tallying a single river and more like balancing the accounts of a city's water supply, with pipes that loop back on themselves.

The OSNAP programme (Overturning in the Subpolar North Atlantic Program) was designed for this geography. It does not rely on one straight line, but on two coordinated transects: one from Labrador toward Greenland, and another from Greenland toward Scotland, spanning the entrances and exits through which the subpolar Atlantic exchanges water with the rest of the ocean. The lines are studded with moorings, like RAPID, but the logic is different. Instead of one basin-wide fence catching the net transport at a fixed latitude, OSNAP aims to capture where water masses are actually transformed—where warm inflows lose heat to the atmosphere, gain or lose freshwater, and become dense enough to slip beneath lighter layers.

The subpolar Atlantic is also where the atmosphere does its most forceful work on the ocean. Winter storms tear heat out of the sea surface and whip waves high enough to make the ocean feel, from the deck of a ship, like a moving landscape rather than a fluid. These storms deepen the surface mixed layer, pulling water from below into contact with the cold air. That mixing is not a detail; it is the mechanism by which the North Atlantic converts surface water into something dense enough to contribute to the deep limb of the AMOC. The intensity of this interaction is why the region matters for climate: it is a primary place where the ocean gives up heat to the atmosphere, moderating European winters and shaping storm tracks.

For a long time, a particular narrative dominated how people talked about overturning in this region. The Labrador Sea, wedged between Canada and Greenland, was treated as the main factory of deep water. Wintertime *deep convection*—the sinking of surface water after strong cooling—was dramatic there, and it was easy to imagine that the strength of convection acted like a dial controlling the AMOC. That idea seeped into simple explanations: more convection, stronger overturning; less convection, weaker overturning.

Observations have a habit of complicating neat dials. When OSNAP began delivering its first sustained record, results led by Susan Lozier and colleagues (2019) challenged the Labrador-centric picture. The data suggested that much of the overturning variability across OSNAP's lines was not dominated by Labrador Sea convection, but by transformations and currents in the Irminger Sea and Iceland Basin. The claim was not that the Labrador Sea is unimportant—it clearly participates in the production and modification of water masses—but that the subpolar system's variability is distributed across pathways shaped by topography and boundary currents. For climate science, this mattered deeply: it implied that focusing on one dramatic patch of winter convection may miss the true levers controlling overturning from year to year.

OSNAP, like RAPID, is an exercise in engineering endurance. Moorings must survive icing conditions and strong currents. Instruments must be calibrated and cross-checked, then recovered and redeployed on servicing cruises that are themselves constrained by weather windows and geopolitics. Yet OSNAP also illustrates a broader truth about modern oceanography: even a heroic mooring array cannot be everywhere at once. The ocean's variability is three-dimensional. It is not only a matter of transport along a line, but of where and when water changes its identity.

That need—to follow the ocean not only at a boundary but throughout its interior—is what turned autonomous robotics from an intriguing novelty into a pillar of climate observation. The transformation is captured in a single word that now sits quietly in countless scientific papers: Argo.

Argo is a global fleet of profiling floats, each one roughly the size of a person, drifting with the currents like a patient messenger. A typical float follows a simple rhythm. It spends most of its time parked at depth, often around 1000 m, where it drifts with the ocean's slower interior flow. Every 10 days or so, it sinks further—commonly to 2000 m in the core Argo programme—and then begins a controlled rise to the surface. As it ascends, it measures a vertical profile of temperature and salinity. At the surface it phones home via satellite, transmitting its data, then dives again to repeat the cycle.

That cycle turned the ocean from something sampled episodically into something sampled routinely. Before Argo, much of the world's subsurface temperature record came from ships and expendable probes dropped along shipping routes. Coverage was uneven: dense in the North Atlantic and North Pacific, sparse in the Southern Ocean and much of the tropics. Salinity, which is essential for density and thus for thermohaline circulation, was even more poorly observed. Argo provided, for the first time, a near-global, near-uniform dataset of the upper 2000 m, delivered fast enough to be used not only for retrospective studies but for operational forecasting. The achievement is partly technological and partly sociological: a shared international commitment to deploy, maintain, and freely distribute data from thousands of instruments without the promise of immediate local payoff.

A concrete example of Argo's power is almost mundane: it made it possible to say, with confidence, where the planet's excess heat is going. The atmosphere warms quickly, and that is what people feel, but most of the heat trapped by greenhouse gases is stored in the ocean. Quantifying that storage requires temperature measurements that are global and repeated. With Argo, ocean heat content estimates became more robust, with smaller uncertainties and fewer blind spots. This is not a minor bookkeeping exercise. The ocean's heat uptake shapes the pace of surface warming and influences sea level rise through thermal expansion. When climate scientists argue about how fast the world is warming, they are often arguing indirectly about how the ocean is absorbing and redistributing heat.

In the North Atlantic, Argo floats complement moorings in a particularly elegant way. A mooring gives you continuity at a fixed location: it watches the ocean's pulse without blinking. A float gives you spatial context: it shows how temperature and salinity vary across fronts, how eddies transport water properties, and how the stratification changes with season. Together they connect the large-scale circulation to the local physics of density. When a winter storm deepens the mixed layer, Argo profiles show the deepening directly. When a fresh surface lens from sea-ice melt or Greenland runoff caps the water column, resisting mixing, Argo sees the salinity gradient that helps keep lighter water afloat.

The Argo story also includes a telling historical twist. In Homer's *Odyssey*, Argo is not a character but a ship, the vessel that carried Jason and the Argonauts. The name was chosen deliberately, yet the modern Argo fleet is the opposite of a single legendary craft: it is a distributed system whose strength comes from redundancy. If one float dies, another nearby may fill the gap. If a region becomes undersampled, deployments can be shifted. This is a new style of observing, closer to ecological monitoring than to heroic exploration. It depends less on one decisive expedition than on an ongoing background effort that is easy to take for granted until it fails.

Taking a float from concept to ocean reality required a series of small engineering solutions to problems that become merciless at depth. A float must change its buoyancy without propellers, using a pump to move oil into an external bladder, subtly increasing volume and thus decreasing density so it rises. It must carry batteries that survive years of pressure cycles. It must measure temperature to thousandths of a degree and conductivity (from which salinity is derived) with stability that does not drift wildly over time. It must detect when it has reached the surface in heavy seas and establish a satellite connection quickly enough to transmit before waves bury the antenna. If this sounds finicky, it is, and it is precisely this finickiness that makes the achievement so impressive: thousands of times a day, somewhere in the world ocean, an Argo float completes a profile and sends it to the public.

Argo's success has also exposed its own limits, and those limits point directly toward the next observational frontier: the deep ocean and the ocean's chemistry.

The core Argo programme samples to 2000 m, which sounds deep until one remembers that the mean ocean depth is closer to 3700 m, and that vast volumes lie below 2000 m. Those abyssal waters change more slowly, but "slowly" does not mean "never." Over decades, the deep ocean can warm, freshen, and shift its circulation pathways. Those changes matter for sea level and for the long memory of the climate system. They also matter for the AMOC, whose deep limb is, by definition, deep. A monitoring system that misses half the ocean's volume is like a medical test that ignores half the body.

Deep Argo is the response: floats designed to profile to 4000 m, 5000 m, and even 6000 m in some models, reaching the abyssal plains where pressures are hundreds of times atmospheric. David Roemmich, Janet Sherman, and collaborators (2019) described one such approach, Deep SOLO, emphasizing the practical details: pressure housings that resist crushing, hydraulics that work reliably under extreme conditions, and mission software that can be updated remotely. One of the most revealing phrases in accounts of these floats is "bottom detection." A deep float must know when it is approaching the seafloor so it does not drive itself into sediment or become trapped. That requirement sounds obvious, but it captures the engineering reality: the ocean is not an abstract volume; it has terrain, and the instruments must negotiate it with imperfect information.

There is also the ocean's biogeochemistry: oxygen, nutrients, acidity, and biological activity. The thermohaline circulation is built on density, which depends primarily on temperature and salinity, but its climate consequences extend well beyond physics. The ocean absorbs carbon dioxide, stores it at depth, and releases it again where waters upwell. It ventilates the deep sea with oxygen. It controls habitats by moving heat and nutrients. Observing only temperature and salinity tells you about the skeleton of the system, but not the full physiology.

Biogeochemical Argo (often shortened to BGC-Argo) extends the float concept by adding sensors for variables such as oxygen, nitrate, pH, chlorophyll fluorescence, backscatter (a proxy for particles), and irradiance. These are harder measurements than temperature and salinity. Sensors can drift, foul, and respond to pressure and temperature in complex ways. Field calibration is not trivial. Francesca D'Ortenzio and colleagues (2020), describing a Mediterranean pilot array, stressed that scaling up such measurements requires careful quality control and consistent practices. The romantic picture of robotic exploration gives way to something more like laboratory discipline performed at sea: calibration curves, cross-comparisons, and a constant effort to ensure that a trend in the data reflects the ocean, not the instrument.

In 2022, William Owens and Nathalie Zilberman helped frame these expansions under the umbrella of "OneArgo," an integrated vision that treats the observing system as a coordinated whole: enhanced core sampling (including difficult regions like high latitudes and marginal seas), deep profiling to capture the abyss, and biogeochemical measurements to connect physics with carbon and ecosystems. What is striking about this vision is that it redefines what counts as "taking the ocean's pulse." The pulse is no longer just temperature; it is oxygen, acidity, and the distribution of life-supporting nutrients. It is also a commitment to long-term continuity, because carbon budgets and ecosystem shifts do not reveal themselves on a single cruise.

How does all of this connect back to the AMOC? The connection is not a single metric but a web of constraints. OSNAP constrains the overturning and its water-mass transformations in the subpolar North Atlantic. RAPID constrains the transport farther south. Argo fills the interior with repeated profiles that show how density structure evolves across seasons and years. Deep Argo begins to extend that constraint downward into the deep limb itself. BGC-Argo adds information about where waters are being ventilated, where carbon is being stored, and how biological productivity responds to physical changes.

There is a human story behind these systems that is easy to miss because it does not centre on one charismatic discovery. Argo floats are deployed from research vessels, but also from ships of opportunity: commercial vessels, national navies, and sometimes even sailing yachts in remote regions. A deployment can be almost comically simple: a crew member heaves a cylindrical float over the stern, it splashes, rights itself, and disappears. For a moment it looks like litter. Then, perhaps ten days later, it surfaces somewhere else and sends back a profile that slots into a global database used by scientists and forecasters around the world. That quiet chain—from a splash in the open ocean to a climate dataset—is one of the most modern things oceanography has ever done.

It also changes the emotional tone of the science. Moorings and ships invite a particular kind of attachment: named cruises, familiar instruments, the drama of recovery. Floats invite a different feeling, closer to watching a flock: individual units matter less than the pattern they form. Yet when a float survives for years, profiling faithfully through storms and seasons, it becomes hard not to feel a kind of admiration. Each float is a small, stubborn witness to the ocean's variability, doing its work without fanfare.

On a winter day in the subpolar Atlantic, when the air is sharp and the sea seems intent on dismantling anything humans put into it, the idea that the ocean can be monitored continuously can feel almost implausible. And yet the data arrive: profiles of temperature and salinity, hints of fresh lenses and deep mixing, oxygen signals that speak of ventilation, pressure records on moorings that register the slow heave of the deep ocean. The ocean remains vast and partially unknowable, but the blank spaces on the map are no longer purely geographical. They are increasingly filled with time series: the ocean not as a single snapshot, but as an evolving, measured system whose changes can be tracked in the same way one tracks the changing weather.

7.3 Simulating the Abyss

A mooring line gives the ocean a place to speak, and a float gives it the freedom to speak almost anywhere. But even a planet studded with instruments does not automatically yield a prediction. The raw observations arrive as fragments: a temperature profile here, a pressure fluctuation there, a current meter's record that drifts and then snaps back. The ocean is continuous; the measurements are not. Between the data points lies the world that matters most for climate: the three-dimensional flow field that carries heat, salt, and carbon, and that can rearrange itself in ways no single instrument can fully witness.

Turning those fragments into a coherent picture is the job of numerical models, but not in the sense of a purely theoretical simulation sealed off from reality. Modern oceanography has developed a hybrid practice that sits somewhere between weather forecasting and detective work. A model provides the rules—the equations of motion, conservation of heat and salt, and the constraints of coastlines and bottom topography. Observations provide the evidence. The challenge is to find an evolving ocean state that obeys the rules while fitting the evidence as closely as possible, without cheating.

Cheating is not a moral failing; it is a numerical temptation. If a model is allowed to add or remove heat and salt arbitrarily to match observations, it can be made to fit almost anything, but it will not represent the real physics. The strongest forms of model—data fusion insist that the model remains dynamically consistent: currents must be driven by forces the model actually includes, and changes in temperature and salinity must come from advection, mixing, and surface fluxes rather than hidden adjustments. The phrase can sound abstract, but its meaning is concrete. A dynamically consistent ocean state is one in which, if you follow a parcel of water in the model, its temperature and salinity change for reasons the ocean could plausibly have.

One influential example is the ECCO project, short for Estimating the Circulation and Climate of the Ocean. In 2015, Gaël Forget, Jean-

Michel Campin, and collaborators described ECCO version 4 as an integrated framework for global ocean state estimation built on the MIT general circulation model. The ambition is straightforward to state and hard to execute: adjust uncertain inputs—winds, surface heat fluxes, initial conditions, and some model parameters—until the simulated ocean matches a wide collection of observations as well as possible, in a least-squares sense, while never violating the model's governing equations.

The phrase *least-squares* sounds like a classroom method, but it is at the heart of the practical problem. Imagine asking, "What did the ocean look like on 3 March 2009?" You might have satellite measurements of sea surface height and temperature, Argo profiles scattered across basins, and a few moorings with continuous records. Those observations are not perfectly consistent with one another; each has errors, and some represent different aspects of the ocean's state. The sea surface height seen by a satellite depends on heat content (warm water expands), on salinity (fresh water is lighter), on currents (which tilt the surface), and on atmospheric pressure. Argo profiles measure temperature and salinity but only where floats happen to be. Moorings measure time variability at fixed points but can miss eddies passing nearby. A state estimate tries to reconcile these partial glimpses into a single evolving ocean that is, in the model's terms, physically possible.

This reconciliation is not just an aesthetic exercise. It produces fields that can be used to compute quantities that are otherwise difficult to observe directly: the transport of heat across a latitude, the overturning streamfunction in density space, the pathways by which deep waters are ventilated, and the regions where mixing is doing most of the work. It also provides something that observationalists and modellers both crave: a common language. A mooring array might measure the AMOC at 26.5°N as a time series of transport in Sverdrups. A climate model might produce an overturning circulation but with different sampling and different definitions. A state estimate can be sampled the way the moorings sample, making a more honest comparison.

That honesty matters because ocean variability is not a nuisance layered on top of a trend; it is the medium in which the trend must be detected. The RAPID array made this painfully clear by showing that the AMOC can swing strongly over days to seasons. Those swings can mask or mimic long-term change. If a model produces a smooth, gently drifting AMOC time series, it may still be "right" about long-term decline, but it will be wrong about what detection feels like in the real ocean. A smooth model encourages overconfidence: it suggests that a decade-long trend should pop out clearly. A noisy ocean suggests the opposite: the trend might be real and yet remain ambiguous for years.

This mismatch between the ocean's jagged reality and the model's smooth tendencies is one reason the assimilation of observations into models became so central. Weather forecasting learned long ago that it is not enough to have the right equations; you must also start from the right initial conditions, updated constantly. The ocean, with its slower time scales, is not forecast in exactly the same way, but the logic carries over. To know where the ocean is going, it helps to know precisely where it is now, and to ensure that "now" is a state that could genuinely arise from the physical rules.

There is a useful historical anecdote here, one that captures both the promise and the discomfort of this approach. In the early days of the RAPID array, the sudden swings in the AMOC time series produced a kind of vertigo. Were those swings real, or artifacts of measurement? Continuous observing systems are, by nature, complicated. They combine instruments with different failure modes, require calibration, and rely on assumptions about geostrophic balance and reference levels. When the first dramatic fluctuations appeared, the scientific response was not to declare victory but to cross-check relentlessly. This is where state estimates and models became a form of skepticism made operational. If an ECCO-like estimate, constrained by many different data types, could reproduce similar variability at the same latitude, it would increase confidence that the swings reflected ocean dynamics rather than instrumentation quirks. If it could not, it would point back to possible problems in either the observing system or the model's representation of key processes.

Baehr and colleagues (2009) offered a concrete example of this style of comparison by examining how the RAPID time series could be used to evaluate modelled overturning variability, including an ECCO-GODAE state estimate. The details are technical, but the philosophical move is simple: treat the observing system not as an isolated truth and the model as an isolated theory, but as two imperfect witnesses that can interrogate each other. This back-and-forth is how modern climate science hardens into something more reliable than either instrument readings or computer code alone.

The observation set that a project like ECCO draws on is strikingly heterogeneous. Satellite altimeters measure sea surface height with near-global coverage, capturing the ocean's surface slopes and eddy field. Satellites measure sea surface temperature, which is the ocean's skin and the interface through which heat is exchanged with the atmosphere. Increasingly, satellites also estimate sea surface salinity, adding a crucial ingredient for density. Gravity missions such as GRACE detect changes in Earth's mass distribution, which can be used to infer changes in ocean bottom pressure and thus in large-scale mass redistribution. In situ measurements from Argo provide temperature and salinity profiles, while ship-based hydrography (from programmes descended from WOCE and continued in GO-SHIP) provides high-accuracy full-depth sections that anchor the long-term calibration of the global system. Sea ice concentration, winds, and atmospheric reanalyses provide the forcing.

It can feel like overkill to throw so many data streams into a single framework, but the ocean is a system where different variables compensate. A rise in sea surface height could reflect warming (thermal expansion), freshening (lower density), or a change in currents (dynamic topography). Without temperature and salinity profiles, you may misattribute the cause. Without gravity constraints, you may miss mass changes that do not show up clearly in surface height alone. The strength of assimilation lies in the way it forces these stories to agree: a sea level rise that the model attributes to warming must also be compatible with the temperature profiles, and it must be compatible with the surface heat fluxes that would be required to produce that warming.

This insistence on consistency has a deeper payoff for the thermohaline circulation. The AMOC is, at its core, a density-driven overturning: differences in density caused by temperature and salinity. Models can produce an overturning circulation for the wrong reasons. They might mix too much, smoothing out stratification until sinking becomes too easy. They might represent deep convection as a blunt vertical mixing event that happens too readily. They might misplace the delicate balance between fresh Arctic inflows and salty Atlantic waters. Assimilating real temperature and salinity fields helps constrain these tendencies, not perfectly, but enough to expose when a model is drifting into a physically implausible state.

There is also an underappreciated emotional dimension to this work: the discomfort of confronting how much the ocean can do without asking permission. State estimation is a form of humility. It begins with the recognition that neither observations nor models alone are sufficient. Observations are too sparse to resolve every eddy and pathway; models are too idealized to trust blindly. The compromise is powerful but never final. Each new dataset—a denser Argo coverage, a better gravity mission, a refined mooring record—can force the state estimate to change, sometimes in ways that overturn comfortable narratives.

This is where the language of "tipping points" enters, often awkwardly. In popular discussion, a tipping point is a dramatic cliff: one more push and the system falls into a new state. For the AMOC, the concern is that gradual warming and freshening could push the circulation past a threshold beyond which it weakens sharply or collapses. The physics behind this possibility involves feedbacks: for instance, a weakened AMOC might transport less salt northward, making the North Atlantic fresher and lighter, which could further weaken sinking and thus weaken the AMOC again. Such feedbacks can create multiple stable states in simplified models.

Yet detecting proximity to a threshold in the real ocean is extraordinarily hard. The ocean is noisy, and the drivers are multiple. A decade of weakening could be a trend or a fluctuation. A sudden dip could be a weather-driven event or a sign that a stabilizing mechanism is failing. This is precisely where model—data synthesis earns its keep. A state estimate can provide a physically consistent reconstruction of the recent past—a *hindcast*—that can be used to test whether a model's mechanisms resemble the ocean's. Does the model weaken for the same reasons the reconstructed ocean weakens? Does it show similar sensitivity to freshwater anomalies? Does it reproduce the observed relationship between winds, boundary currents, and overturning strength? If not, then its projection of a future threshold is less credible, even if the projection seems dramatic.

Climate model ensembles, such as those coordinated under CMIP6, are often used to explore possible AMOC futures under different greenhouse gas scenarios. These ensembles produce a spread of outcomes: some show substantial weakening over the twenty-first century, others show more moderate change. The spread reflects both genuine uncertainty and differences in how models represent key processes such as mixing, deep convection, and freshwater fluxes. Observations and state estimates narrow this uncertainty not by picking a single "winner" model, but by ruling out behaviours that are inconsistent with the observed present-day system. A model that cannot reproduce the amplitude and timing of subpolar overturning variability, or that requires implausible surface flux adjustments to match the observed temperature field, earns less trust when it predicts a threshold.

There is another, more practical way to think about the value of simulation: it creates a laboratory the ocean will not provide. No one can run the twentieth century again with a slightly different wind pattern. No one can add a controlled amount of freshwater to the North Atlantic and watch what happens, except in a model. But the laboratory is only useful if its baseline state resembles the real ocean. Assimilation and hindcasting are what make that resemblance testable.

ECCO's emphasis on reproducibility and open availability adds a further twist that is cultural rather than physical. The oceanographic community has not always been an open-data culture. Ship expeditions were expensive, and data could be guarded. Models were bespoke and hard to reproduce. Systems like ECCO helped normalize the idea that a state estimate should be something another researcher can download, examine, and re-run with community tools. That openness does not eliminate disagreement, but it changes its texture. Debates can focus more on physical interpretation and less on inaccessible methodology. For a subject as politically charged as climate change, this matters: credibility grows when the pipeline from observation to inference is visible.

None of this should be mistaken for the claim that we can "solve" the ocean. The abyss remains, in many ways, the most alien part of Earth's climate system: cold, dark, and difficult to observe, with time scales that stretch beyond a human career. Simulation does not tame that alienness; it gives it a form that can be manipulated and questioned. It allows scientists to ask not only "what is the AMOC doing?" but "what would have to be true for it to behave differently?" The answer is rarely a single lever. It is a balance of winds and buoyancy forcing, of salinity and temperature, of mixing and boundary currents, of the subtle way topography guides deep flow like a hidden set of walls.

A useful image is to think of the ocean state estimate as a kind of negotiated truce between what we have measured and what we believe about physics. When the truce holds, it yields a coherent picture that can be probed for mechanisms and sensitivities. When it fails—when the model must be pushed too hard to fit the data, or when the data refuse to be reconciled—that failure is itself information. It tells us where the observing system is insufficient, where the model physics are missing something, or where the ocean is doing something we did not anticipate.

On a map, the abyss is an expanse of uniform blue. In a state estimate, it becomes a moving architecture: deep western boundary currents hugging continental slopes, dense overflows spilling across sills, slow interior upwelling spread thinly across basins, and eddies stirring the boundaries between water masses. It is not an artistic flourish. Those deep motions are part of how the ocean stores heat and carbon, and part of how the AMOC can change without anyone at the surface noticing right away. The most unsettling possibility is not a spectacular, instant collapse, but a deep rearrangement that continues quietly for years and only later announces itself through altered sea level patterns, shifting storm tracks, and a changed distribution of ocean heat.

The point of simulating the abyss is not to replace reality with computation, but to make reality legible enough that the future can be discussed without hand-waving. When the model and the observations align, the alignment is not a guarantee. It is a hard-won statement that, for a time and within uncertainties, the ocean's vast machinery is behaving in ways our physics can track. In a warming world, that capacity to track—to distinguish a wobble from a drift, and a drift from an approach toward a threshold—is the difference between guessing and understanding.

Chapter 8

A World Transformed

While the planet warms, a collapse of the Atlantic circulation promises a cruel paradox: a localized deep freeze for Europe contrasted with intensified heating in the Southern Hemisphere. This chapter traces the physical and chemical shockwaves of a stalled conveyor belt, moving from the disruption of ancient monsoon cycles to the physics of rising seas and the dangerous saturation of the ocean's carbon sponge.

8.1 The Cold Shadow

There is a particular kind of misunderstanding that thrives in a warming world: the idea that every place should get warmer in the same way, at the same pace, year after year. The ocean makes a mockery of that expectation. By moving heat like a vast conveyor belt—and by sometimes faltering—the ocean can create a cold region inside a warming planet, a shadow cast not by the Sun but by circulation.

A useful image is to picture the Atlantic as a giant heat pipeline. The tropics receive an enormous surplus of solar energy. Winds and currents do not let that heat sit still; they export it. One branch of that export runs north through the Atlantic, where warm surface waters release heat to the atmosphere. The air over the North Atlantic and western Europe is unusually mild for its latitude partly because the ocean is constantly paying out heat like a pension: a steady transfer from seawater to sky. If that payout shrinks, the atmosphere above it feels the loss quickly, even if greenhouse gases are still rising in the background.

That loss does not simply vanish. Energy is conserved. When less heat is delivered to the North Atlantic and released to the air, more of it remains elsewhere in the ocean-atmosphere system. The most dramatic expression of that redistribution is the *bipolar seesaw*: a tendency for the Northern Hemisphere, especially around the North Atlantic, to cool (or warm less) while the Southern Hemisphere warms (or warms more), and vice versa. The seesaw is not a neat mechanical lever with a fixed pivot; it is a statistical pattern that emerges from how ocean circulation links the hemispheres. But it is strong enough that it has left fingerprints in ice cores, marine sediments, and the geometry of rainfall belts.

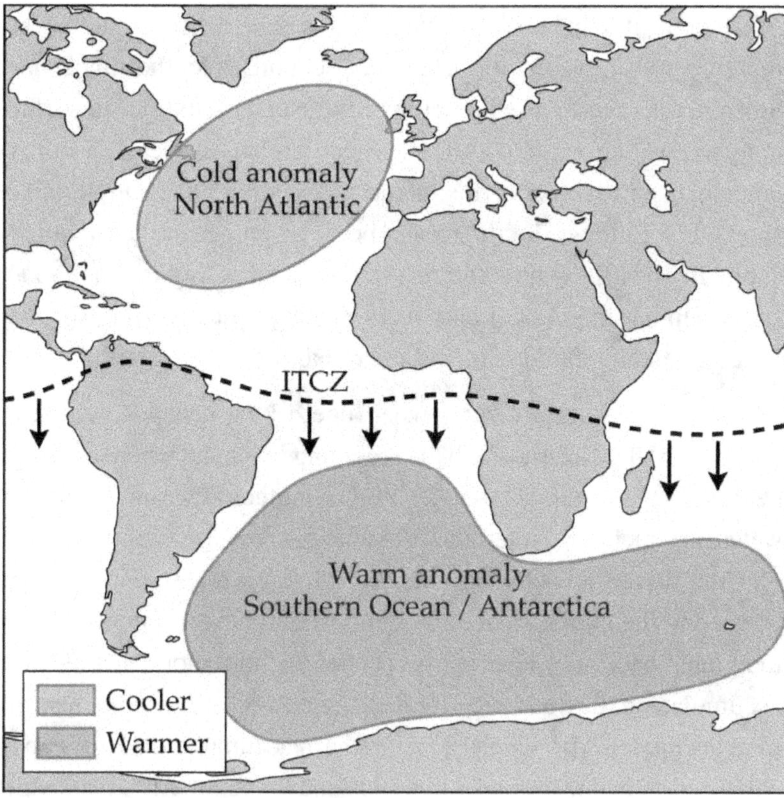

The North Atlantic patch of relative cooling has a name in contemporary climate discussions: the *warming hole*. It is "warming" only in the sense that the planet is warming overall; locally, the surface waters in parts of the subpolar North Atlantic have warmed less than much of the global ocean, and in some analyses even cooled over particular periods. That patchiness can arise from several causes, but a weakening of the Atlantic's overturning circulation is one of the mechanisms that can produce the right pattern: less warm water carried northward, less heat released to the air, cooler surface waters persisting longer into the year.

To understand why the cooling has such leverage over weather and livelihoods, it helps to notice where it sits. The North Atlantic is not merely a body of water; it is a boundary condition for the atmosphere. Storm tracks feed on temperature contrasts, and the ocean surface temperature sets those contrasts. A cooler-than-expected North Atlantic can shift where storms intensify and where they dump rain or snow. For Europe, the difference between an ocean that is quietly supplying heat in winter and one that is withholding it is not an abstract number in a model output; it translates into growing seasons, heating demand, flood risk, and the reliability of harvests.

The seesaw becomes clearer when one asks a deceptively simple question: if the North Atlantic is receiving less heat, where does that heat go? Part of the answer lies in the Southern Ocean, the ring of water that encircles Antarctica. It is the planet's most vigorous zone of wind-driven upwelling and a crucial gateway between the deep ocean and the atmosphere. When the Atlantic overturning weakens, the global pattern of heat transport changes, and more heat tends to accumulate in the Southern Hemisphere ocean. In past climates, and in many model experiments, this southern accumulation shows up as enhanced warming around Antarctica and the Southern Ocean while the North Atlantic cools. The hemispheres do not behave like twins; they behave like partners tied together by an exchange of heat and freshwater.

A historical anecdote makes this feel less theoretical. During the last ice age, the climate lurched through abrupt events that unfolded over decades to centuries—fast by geological standards, slow by the tempo of headlines but quick enough to reshape ecosystems and human possibilities. Greenland ice cores record dramatic warmings (some in a matter of decades) followed by slower coolings. Meanwhile, Antarctic ice cores often show a different rhythm: more gradual changes, sometimes warming while Greenland was cold. This out-of-step pattern was one of the clues that led scientists to articulate the bipolar seesaw in the first place. The idea is not that Antarctica responds instantly, but that the ocean's reorganization tends to redistribute heat between hemispheres in a way that leaves a characteristic anti-correlation.

One famous family of episodes in the North Atlantic are the *Heinrich events*: times when enormous numbers of icebergs broke from ice sheets and drifted into the North Atlantic, dropping rock debris onto the seafloor as they melted. Those icebergs did more than litter the ocean floor. They freshened the surface waters. Fresh water is lighter than salty water, so a surface freshening can act like a lid, inhibiting the sinking of dense water that helps drive the overturning circulation. When that sinking is suppressed, the northward flow of warm surface water weakens. The North Atlantic cools. And, consistent with the seesaw, the Southern Hemisphere tends to warm, with knock-on effects that can include shifts in sea ice and changes in the Southern Ocean's ventilation.

That story matters now not because we expect a neat replay of an ice-age iceberg armada, but because it demonstrates a principle: the overturning circulation can change state, and when it does, the climate response is global yet uneven. The North Atlantic is not the only actor, but it is an outsized one. The Atlantic is narrow enough, and the high-latitude North Atlantic is active enough in exchanging heat with the air, that changes in its circulation readily imprint on temperatures felt on land.

The seesaw is also expressed not only in temperature but in rain. Earth's major tropical rain belt is organized around the *Intertropical Convergence Zone* (ITCZ), where trade winds from the two hemispheres meet and air rises, condensing moisture into towering convective storms. The ITCZ is not fixed at the equator; it migrates north and south with the seasons, following the zone of maximum heating. But it also responds to imbalances between the hemispheres. If the Northern Hemisphere cools relative to the Southern Hemisphere, the atmosphere tends to transport energy northward to compensate, and one way it adjusts is by shifting the ITCZ toward the warmer hemisphere—often southward when the north cools.

This is where a circulation change in the North Atlantic can end up written into the fate of crops thousands of kilometers away. A southward shift of the ITCZ can reduce rainfall over the Sahel, the semi-arid belt just south of the Sahara. The Sahel's rains are famously variable, swinging between drought and relative greening. Many factors drive that variability, including sea surface temperatures in multiple basins and aerosols, but the underlying physics gives the North Atlantic a route of influence: cool the North Atlantic, tilt the hemispheric temperature balance, nudge the tropical rain belt south. The result is not a gentle change in average drizzle; it can be a shift in whether the rainy season arrives on time, how long it lasts, and whether pasturelands can sustain herds.

The social stakes of such shifts are plain in modern history. The Sahel droughts of the late twentieth century were devastating, contributing to famine, displacement, and political stress across several countries. Those droughts cannot be attributed to a single ocean mechanism. Yet they serve as a sober reminder that rainfall belts are not just lines on meteorological charts; they are infrastructure as real as dams and roads, except built from moving air and evaporating water. When the ITCZ shifts, the infrastructure moves.

Europe, too, sits under a kind of atmospheric infrastructure: the storm track of the North Atlantic. The mildness of western Europe is often described in casual conversation as "the Gulf Stream," a shorthand that blends ocean and atmosphere. The reality is more layered, involving westerly winds, ocean heat release, and the geography of land and sea. Still, the shorthand points to a truth: western Europe's climate is unusually sensitive to the state of the North Atlantic. If the ocean surface cools, it does not simply subtract a degree or two from a thermometer. It reshapes pressure patterns and humidity gradients that determine the character of winters.

Agriculture is where those shifts become intimate. Consider the difference between a winter that repeatedly hovers just above freezing and one that repeatedly dips below. For perennial crops, fruit trees, and winter cereals, that difference can determine frost damage, pest survival, and the timing of flowering. A colder North Atlantic can also influence the likelihood of blocking patterns that lock weather into place, raising the risk of prolonged cold spells in some regions even as the global mean temperature rises. A "cold shadow" does not abolish global warming; it complicates it, producing a world in which the phrase "warmer on average" coexists with "colder here, sometimes, in ways that matter."

The seesaw also highlights an uncomfortable possibility: a partial or regional cooling can be misread as evidence against climate change. Human perception is local and immediate. If a few harsh winters cluster in memory, or if a cool patch appears on maps, it can seem like a contradiction. Yet a cooling North Atlantic in a warming world is not a rebuttal to greenhouse physics; it is a consequence of adding greenhouse gases to a planet whose heat is redistributed by moving water. The atmosphere traps more heat; the ocean decides, to a surprising degree, where that heat shows up first.

Behind the scenes of the cold shadow lies a set of feedbacks that can amplify change once it starts. One of the most discussed is the salt-advection feedback. The Atlantic overturning circulation carries not only heat but salt. If the circulation weakens, the pattern of salt transport shifts. In some configurations, a weakened circulation brings less salty water into the regions where dense water forms. Less salt means lower density; lower density makes sinking harder; sinking becomes weaker still. This is not a guarantee of runaway collapse, but it is a mechanism by which the system can possess thresholds: push it gently for a while, and then it flips.

Henry Stommel, in a remarkably compact paper in 1961, demonstrated with a conceptual model that a thermohaline circulation (a circulation driven by heat and salt) can have more than one stable state under the same external conditions. That was a profound idea. It meant that the ocean could, in principle, jump between regimes rather than slide smoothly along a single track. Modern climate models are vastly more complex than Stommel's thought experiment, but the conceptual lesson endures: the ocean's overturning is not necessarily a dimmer switch. It can behave like a system with memory and reluctance, resisting change until it does not.

The cold shadow, then, is not merely a patch of blue on a temperature anomaly map. It is a sign that the ocean's internal plumbing is changing, and that the consequences are distributed unevenly across the globe. In the North Atlantic region, a weaker delivery of oceanic heat can cool the sea surface and modify weather patterns. In the Southern Hemisphere, heat tends to build, with implications for sea ice, ecosystems, and the stability of ice sheets that rest on bedrock below sea level. In the tropics, rainfall belts can shift, not because the tropics are directly connected to the North Atlantic by a narrow current, but because the entire atmosphere responds to hemispheric imbalance.

There is also a more philosophical sting in the seesaw. It undermines the comforting notion that climate change is a uniform background trend on top of which "normal" variability plays. Instead, variability and trend can become entangled. A circulation change alters variability itself: where storms travel, where rain falls, where the ocean stores heat. The cold shadow is the climate system reminding us that it is a coupled machine, not a simple thermostat.

Walk along a windy shore in Ireland, or watch winter waves batter a seawall in Brittany, and it is easy to forget that this violence is partly powered by invisible gradients of temperature between ocean and land, tropics and poles. Those gradients are being redrawn. A weakening Atlantic heat supply can make a coastline feel, at times, more like its latitude suggests it "should" feel. Meanwhile, the extra heat does not disappear; it arrives elsewhere, warming waters that lap at other coasts and stressing other ecosystems.

The bipolar seesaw is not destiny, but it is a warning against thinking locally about a global engine. The ocean can create cold where we expect warmth, drought where we expect rain, and storms where we expect quiet, by shifting the pathways along which heat travels. The shadow it casts over the North Atlantic is cold precisely because the rest of the world is not allowed to be simple.

8.2 The Slumping Wall of Water

Sea-level rise is often pictured as a slow bath filling: ice melts, the ocean expands as it warms, coastlines gradually surrender. That picture is not wrong, but it is incomplete in a way that matters for anyone living along the Atlantic rim. The ocean does not have a perfectly level surface, even on calm days. It has hills and valleys measured in tens of centimeters to more than a meter, sculpted by winds and currents. Those slopes are usually invisible to the eye, spread over hundreds of kilometers, but they are as real as a mountain pass. When a major current weakens, some of those hills subside and the water redistributes. Along a coastline, that redistribution can look like a sudden extra pulse of sea-level rise that has little to do with how much ice melted that year.

The Gulf Stream is the best-known Atlantic current, partly because it appears in weather lore, partly because it is dramatic: a river of warm water racing northward along the American continental slope before veering east. Less widely appreciated is that its speed is tied to a sea-surface tilt. Fast currents are not simply water "choosing" to run quickly; they are flows balanced against forces. In the open ocean, away from shore and the friction of shallow water, the most important balance is between the sideways pull of pressure and the apparent deflection caused by Earth's rotation. Oceanographers call this *geostrophic balance*.

A pressure gradient in the ocean is often just a slope in the sea surface. If the surface is slightly higher in one place than another, there is slightly more water above, hence slightly higher pressure below. Water accelerates away from high pressure toward low pressure. Earth's rotation then deflects that moving water: to the right in the Northern Hemisphere, to the left in the Southern. In a geostrophic current, the deflection continually turns the flow so that, instead of running down the slope, it runs *along* the slope. The ocean behaves like a puck sliding on a rotating table, perpetually pulled sideways. The result is counterintuitive: a fast current tends to sit next to a tilted sea surface, with "high" water on one side and "low" water on the other.

Along the U.S. East Coast, the Gulf Stream's tilt acts like a dynamic dam. It helps keep the coastal sea surface slightly lower than it would otherwise be, because the current's structure and speed maintain a pressure field that "leans" the sea surface away from the coast. This is not a seawall you can photograph, and it does not protect against storms in the way a barrier might. It is a background condition, a baseline that influences how high the ocean sits before winds pile extra water on top. If the Gulf Stream slows, the tilt relaxes. Water that was, in effect, held offshore slumps back toward the coast. The coastline experiences a rise in sea level that is driven by circulation itself, often called *dynamic sea level*.

A simple way to feel the physics is to imagine walking across a gently sloped floor. If you place a ball on it, the ball rolls downhill. Now imagine the same floor on a rotating turntable. If you try to roll the ball straight down the slope, the rotation pushes it sideways so that it curves and can end up traveling along a contour instead. That sideways push is the Coriolis effect, and the contour is the sea-surface height line. In the ocean, a strong current is like the ball in steady motion: the slope is still there, but the flow runs along it. Reduce the speed and the balance changes. The "dam" leaks, not through a hole, but through a change in the whole pressure field.

The most unsettling feature of this mechanism is timing. Thermal expansion and melting ice contribute to sea-level rise on long timescales; their effects are cumulative and global. Dynamic sea-level changes can occur much faster and can be strongly regional. A weakening of the Atlantic's overturning circulation—the large-scale system that helps sustain the Gulf Stream and its extensions—does not raise sea level equally everywhere. It tends to raise sea level along the western boundary of the North Atlantic, including much of the northeastern United States and Atlantic Canada, because that is where the current's pressure gradient is most tightly tied to the coast.

The idea that circulation can move sea level around has been part of physical oceanography for a long time, but it came uncomfortably close to everyday experience in 2009–2010. Along parts of the U.S. Northeast coast, tide gauges recorded an unusually large and persistent sea-level anomaly: the ocean sat several centimeters higher than expected for months, contributing to a rash of "nuisance" flooding and amplifying the impacts of storms. This was not the kind of single-day storm surge that makes headlines and then subsides; it was a background elevation, a higher starting point. Later analyses linked this event to a pronounced downturn in Atlantic circulation indices and to atmospheric conditions (including a strongly negative North

Atlantic Oscillation) that altered winds and pressure. The point is not that one winter tells the whole story of the future, but that the ocean demonstrated a capacity for abrupt-seeming coastal consequences without waiting for the slow arithmetic of global mean sea-level rise.

Those centimeters matter. Coastal flooding is threshold-driven. A street floods when water crosses a curb; a subway entrance floods when water tops a stairwell lip. Raise the baseline by even 5–10 cm and the number of high-tide floods can jump, because the distribution of tide heights has a fat tail: there are many more "almost" floods than "major" floods. City engineers and residents often experience this not as a smooth trend but as a sudden increase in how often salt water appears where it did not before.

One might reasonably ask whether a current that flows offshore can really affect water right up against the coastline. The answer lies in how the Gulf Stream hugs the edge of the continental shelf. The shelf is shallow, rougher, and frictional; the deep ocean is smooth by comparison. The Gulf Stream runs along the shelf break, and its pressure field is felt across the shelf and into coastal waters. The height difference maintained by the current is not a sharp step but a broad tilt that can extend over large distances. When the current changes, the adjustment propagates through the ocean in ways that can be rapid, carried by waves and by the rebalancing of pressure fields.

There is also a second strand to the story: winds. Alongshore winds can pile water against a coast through a process called *Ekman transport*, in which the surface layer of the ocean moves at an angle to the wind because of rotation and friction. Persistent winds can therefore raise or lower coastal sea level by shifting surface waters toward or away from shore. In the real world, dynamic sea level reflects a combination of current-driven pressure gradients and wind-driven setup. What makes the Gulf Stream and the larger Atlantic circulation especially important is that they provide a structured, semi-persistent pattern onto which storms and seasonal winds add their own pulses. When that background pattern changes, every storm surge rides on a different pedestal.

The overturning circulation adds depth to the drama. The Gulf Stream is often described as wind-driven, and indeed the winds over the subtropical Atlantic help set up the broader gyre circulation. But the northward flow of warm surface water and the southward return flow at depth are part of a larger system that involves density differences created by temperature and salinity. When high-latitude waters become dense and sink, they help draw surface waters northward to replace them. If that sinking weakens, the whole circulation can rearrange: not necessarily switching off the Gulf Stream overnight, but altering its strength, path, and the sea-level slope it maintains.

A useful mental picture is to think of the Atlantic basin as having a gently tilted surface: lower along the American coast, higher toward the mid-basin and across toward Europe, with the precise shape set by currents. The "tilt" is the pressure gradient that, together with rotation, supports the flow. Reduce the flow and the tilt relaxes; water redistributes; the western boundary becomes higher. The coast does not need to wait for Greenland's ice to melt to feel this effect. It is a redistribution of water already in the basin.

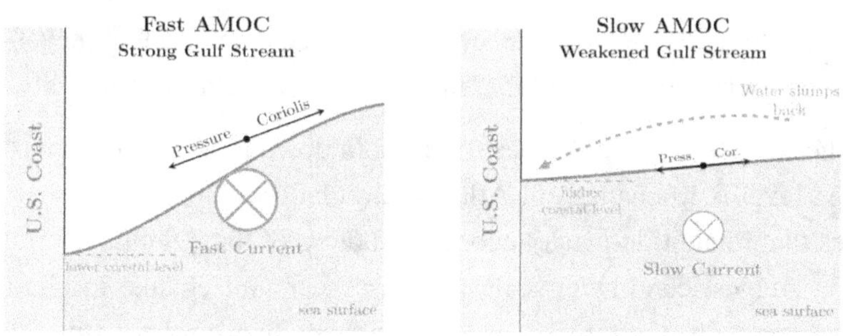

This schematic exaggerates the geometry, but the message is faithful: a fast boundary current is associated with a stronger cross-stream sea-surface slope; weaken the current and the slope flattens, leaving relatively more water on the coastal side. Real coastlines add complications. The continental shelf can trap coastal currents; estuaries and bays can resonate; local winds can dominate during storms. Yet the large-scale pressure field is still the backdrop against which all of that happens.

For communities, the practical question is not whether dynamic sea level exists, but how it combines with other changes. A higher baseline sea level increases the reach of storm surges. It also affects groundwater: salt water can intrude into coastal aquifers from below, and drainage systems that rely on gravity can fail more often when the sea pushes back. Even without a headline-grabbing hurricane, a persistently elevated coastal sea level can cause chronic flooding on sunny days, the kind that rusts car frames, kills roadside trees, and forces businesses to install temporary barriers as routine.

The Atlantic coast is not uniform in its vulnerability. The Mid-Atlantic and Northeast have many low-lying neighborhoods built on filled land, much of it developed when sea level was lower and nuisance flooding rare. The coast is also subsiding in places for geological reasons, which adds to relative sea-level rise. Dynamic changes are an extra term in an already crowded equation. A few centimeters from circulation, added to a few centimeters from thermal expansion, added to local land subsidence, can produce a decade that feels like an acceleration even if global mean sea level is climbing steadily.

The science community has tried to pin down how sensitive coastal sea level is to changes in Atlantic circulation. Observational studies that relate tide-gauge records to indices of the Atlantic overturning suggest centimeter-scale responses per unit change in circulation strength, with the exact number depending on the method, the timescale, and the location. Models, too, show a coherent pattern: weaken the overturning and sea level rises along the western North Atlantic boundary while falling or rising less elsewhere in the basin. International assessments such as the IPCC have highlighted this pathway as a plausible contributor to regional sea-level change, precisely because it does not require an extreme melt event to matter.

A lingering discomfort remains, and it is worth acknowledging openly: measuring changes in the overturning circulation and attributing coastal sea-level anomalies to them is hard. The ocean is noisy. Winds, atmospheric pressure, river discharge, and internal variability all leave signatures on sea level. Direct measurements of the overturning at key latitudes are relatively recent. This is why single events, like the 2009–2010 anomaly, are used not as definitive proof but as demonstrations that the mechanism can operate at societally relevant magnitudes and timescales. The value of such events is almost pedagogical: they show that the ocean can rearrange itself quickly enough to surprise city planners.

The "slumping wall of water" phrase captures the feeling of it: not a new flood poured in from outside, but a redistribution of what was already there, released from a dynamic constraint. The wall is built from pressure gradients and rotation; its mortar is the speed of a current. When that speed falters, the wall relaxes. Along the coast, the change is not a subtlety measured only by satellites and specialists; it is water over a bulkhead, salt on a road, a higher high tide that turns an inconvenience into a disruption.

Stand on an Atlantic pier at high tide and watch the waterline creep up the pilings. The ocean looks level because our eyes are not made for gradients spread across a thousand kilometers. Yet the height of that water at your feet is connected, through geostrophic balance, to the speed of a current offshore and the density structure of the basin. That connection is easy to miss precisely because it is so elegant: the coast is tied to the open ocean not only by waves and storms, but by the shape of the sea surface itself.

8.3 Closing the Carbon Vault

The ocean has been doing humanity a quiet favor. Every ton of carbon dioxide released from a tailpipe or smokestack does not stay in the air. A large fraction is taken up by seawater, and a great deal of the excess heat trapped by greenhouse gases ends up in the ocean as well. Without that buffering, atmospheric warming would have sprinted ahead of what has already unfolded.

But the buffer is not a bottomless pit, and it is not a simple sponge. The ocean hides carbon and heat by moving them downward and sideways, away from the surface that exchanges gases and warmth with the atmosphere. The machinery that performs that hiding is ventilation: the process by which surface waters are carried into the interior and deep ocean. When the overturning circulation is vigorous, the ocean has an efficient elevator. When it slows, the elevator becomes unreliable. The carbon and heat do not vanish; they linger nearer the surface, where they can re-emerge into the air and where they can do damage to ecosystems.

It helps to be clear about what "sequester" means here. The ocean takes up carbon dioxide from the atmosphere because of chemistry: CO_2 dissolves in water, and then reacts to form bicarbonate and carbonate ions. This chemistry allows seawater to hold far more carbon than it could if CO_2 simply dissolved and stayed as a gas. The ocean also takes up heat because water has a large heat capacity; it can absorb a great deal of energy for a modest rise in temperature. Yet neither carbon nor heat is truly removed from the climate system merely by crossing the air–sea boundary. As long as it remains near the surface, it can be returned to the atmosphere by changes in temperature, winds, or mixing. Long-term storage depends on transfer into the ocean interior, where exchange with the air is slow.

Imagine the surface ocean as a thin skin compared with the whole depth of the sea. Most marine life, most ships, and almost all of our measurements live in that skin, but it is only a small fraction of the ocean's volume. For carbon and heat to be stored for centuries, they must be carried below the skin and kept there. The most direct route is deep water formation: places where surface water becomes dense enough to sink, ventilating the abyss and pulling surface properties downward like a stamp imprinting wax.

The North Atlantic is one of the planet's great stamping grounds. In cold, windy regions near Greenland, Iceland, and the Labrador and Nordic Seas, surface waters can lose heat rapidly to the atmosphere. Sea ice formation can also concentrate salt in the remaining liquid water, making it denser. When the mixture of cold and saltiness makes the water heavy enough, it sinks and spreads at depth as North Atlantic Deep Water. This is not just a local process. It is a major pathway through which the ocean interior is renewed, and it is one reason the North Atlantic has stored a disproportionately large share of the anthropogenic carbon taken up by the global ocean.

There is an intuitive logic to that disproportion. If you want to hide something, you need a trapdoor. Much of the ocean is stratified: lighter water sits on heavier water, like oil over vinegar. Stratification inhibits vertical exchange. In most places, carbon and heat absorbed at the surface will be mixed down only slowly, through eddies, turbulence, and the occasional deep winter mixing event. Where deep water forms, the trapdoor opens wide. Surface water can be injected into the interior in large volumes, carrying with it dissolved CO_2 and excess heat.

Now consider what happens when the overturning circulation weakens. Sinking in the North Atlantic becomes less frequent or less deep. The trapdoor narrows. The ocean can still take up carbon at the surface, but it becomes harder to export that carbon into the deep interior. The surface layer becomes more saturated, in a practical sense: not because the chemistry has reached a hard limit, but because the residence time near the surface increases. In a slower system, the ocean's carbon intake can decline even if atmospheric CO_2 remains high, because the surface ocean is not being refreshed as effectively by lower-carbon waters from below.

Heat behaves similarly, with an extra twist. The ocean's ability to absorb heat depends on how quickly heat is moved away from the surface. When the upper ocean warms, it becomes more stratified: warm water is lighter, so it sits more stably atop cooler, denser water. That makes it harder for wind and convection to mix heat downward. A slowing overturning circulation can intensify this surface trapping in some regions, leaving more heat in the upper layers where it affects weather patterns, marine heatwaves, and ice melt at the edges of ice sheets. The deep ocean, paradoxically, may warm more slowly even while the surface experiences sharper extremes.

This is why the circulation is not merely a delivery system for climate impacts; it is also part of the planet's carbon accounting. A weakening of the overturning can turn the ocean from a generous sink into a less reliable one. Earth system model studies that experiment with slower overturning generally find a net reduction in the ocean's uptake of anthropogenic CO_2 over time. There are competing processes: changes in biology, nutrient supply, and regional chemistry can push in different directions. But the broad conclusion is uncomfortable: if the ocean's ventilation engine slows, the atmosphere is left holding more of our emissions, and warming accelerates.

The chemistry adds a humanly tangible consequence: ocean acidification. When CO_2 dissolves in seawater, it forms carbonic acid, which releases hydrogen ions and lowers pH. The shift is modest in absolute numbers but large in chemical terms, and it reduces the availability of carbonate ions that organisms need to build shells and skeletons made of calcium carbonate. Long-term observations in the North Atlantic, including time-series sites that have become oceanographic equivalents of weather stations, have recorded declining pH and declining saturation states for minerals such as aragonite. Those trends track the ocean's uptake of anthropogenic CO_2 as faithfully as a receipt tracks a purchase.

The North Atlantic's special role makes the acidification story especially poignant there. The same processes that ventilate the deep ocean and draw down CO_2 also imprint the surface with strong variability: winters that mix deeper, years when currents shift, patches where biology draws down CO_2 in spring blooms. Against that noisy background, scientists have worked for decades to detect the persistent signal of human emissions. The signal is now clear enough that it has left the realm of subtle statistics and entered the lived reality of industries. Shellfish hatcheries, for example, have had to contend with episodes of corrosive seawater that can interfere with the early life stages of oysters and other species. The causes of those episodes are not always directly tied to the overturning circulation—local upwelling and coastal processes often dominate—but the broader point remains: carbon absorbed by the ocean changes the ocean, and the pace and pattern of absorption depend on circulation.

A historical analogue makes the vulnerability of the "carbon vault" more vivid. During the last deglaciation, when Earth emerged from the last ice age, atmospheric CO_2 rose by roughly 80 parts per million over several thousand years. Ice cores show that this rise was not perfectly smooth; there were periods of faster increase. Multiple lines of evidence suggest that changes in Southern Ocean ventilation played a role: when deep waters in the Southern Ocean were more effectively brought to the surface, carbon stored in the deep ocean could be released to the atmosphere. At the same time, the Northern Hemisphere experienced abrupt climate events linked to changes in Atlantic overturning. The details remain an active area of research, but the overarching lesson is sobering: the ocean can act as a reservoir that, under different circulation states, either locks carbon away or allows it to leak out.

This is not a claim that today's ocean will suddenly belch out centuries of stored anthropogenic carbon in a single dramatic exhalation. The modern ocean is taking up carbon overall, and the chemistry of anthropogenic carbon differs from the natural carbon cycle in important ways. The lesson is subtler and, in some ways, more worrying: the ocean's role as a sink is contingent on circulation. Change the circulation and you change the partitioning of carbon between ocean and air, and you change it on timescales that matter for policy and planning.

Biology complicates the picture further, and it does so in a way that is easy to underestimate. The ocean's carbon storage is often divided into a *solubility pump* and a *biological pump*. The solubility pump is driven by physics and chemistry: cold water can hold more dissolved CO_2 than warm water, so high latitudes tend to absorb CO_2, and if that water sinks, it carries carbon to depth. The biological pump is driven by life: phytoplankton in sunlit waters use CO_2 to build organic matter; some of that organic matter sinks as particles or is transported downward in other forms, carrying carbon into the interior where it can be stored for years to centuries before returning to the surface.

Overturning circulation influences both pumps. By ventilating the deep ocean, it affects how quickly surface waters are replenished with nutrients such as nitrate and phosphate. Those nutrients fertilize phytoplankton, especially in regions where sunlight is abundant but nutrients are scarce. A slowdown in overturning can, in many models, increase the sequestration of nutrients into the deep ocean, leaving the surface more nutrient-poor over long timescales. That can reduce global marine net primary production, the base of the ocean food web. A change that begins as a physical reorganization can thus echo into fisheries, biodiversity, and food security.

There is a cruel symmetry here. The same circulation that helps hide anthropogenic carbon by transporting it downward also helps maintain a nutrient supply that supports surface productivity. If the circulation falters, less carbon may be stored and less biological production may occur. The ocean becomes both a less effective sink and a less generous provider.

The heat side of the "vault" has its own feedbacks. The deep ocean is the planet's great heat reservoir, and the transfer of heat into it is one reason surface warming has been moderated. When less heat is transported downward, more remains near the surface, which can intensify stratification, which can further reduce mixing, a self-reinforcing loop in some regions. Meanwhile, where heat does enter the ocean, it can be delivered to vulnerable places. Warm water at depth can erode ice shelves from below, particularly around Antarctica, where floating shelves buttress land ice. In that sense, the ocean's heat storage is not automatically benign. Storing heat in the wrong place can accelerate sea-level rise by destabilizing ice.

The earlier idea of a "cold shadow" in the North Atlantic can coexist with this heat problem because the ocean redistributes heat rather than simply absorbing it uniformly. A weakened overturning can cool parts of the North Atlantic surface while the Southern Ocean and other regions warm more. The vault is not closing everywhere in the same way. Some regions may continue to take up heat and carbon efficiently; others may stall. The planet cares about the total, but societies experience the pattern.

One might wonder whether the ocean could compensate for a weaker Atlantic by ventilating elsewhere. The Southern Ocean is indeed a powerful ventilator, and it plays a central role in the global overturning circulation. Yet compensation is not guaranteed to preserve the same carbon and heat uptake. Ventilation that brings deep waters *to* the surface can release natural carbon; ventilation that sends surface waters *down* can store anthropogenic carbon. The direction and structure of these pathways matter. A rearranged system can be both more ventilated in one sense and less effective as a sink in another.

The phrase “closing the carbon vault” is deliberately paradoxical. It does not mean the ocean stops interacting with the atmosphere. It means the ocean becomes less able to lock away what it absorbs. The surface continues to exchange gases and heat with the air, but the interior becomes less connected to that exchange. Carbon that would have been transported to depth lingers near the top, acidifying the waters where most marine life lives. Heat that would have been sequestered in the abyss remains in the upper ocean, fueling extremes. The vault door still exists, but it swings more reluctantly.

There is also a moral and political sting. Much of the world has benefited from the ocean’s buffering without paying for it, because the costs are delayed and dispersed. Acidification does not respect national borders; heat stored in one basin can affect storms and ice in another. When the buffer weakens, the bill arrives faster and more locally. Coastal cities that already face the slumping wall of water face a compounding risk: higher baseline sea level, more intense rainfall in a warmer atmosphere, and a changing ocean that can no longer be counted on to quietly swallow a steady share of emissions.

For an individual, the ocean can feel eternal and inexhaustible. The horizon looks the same year after year. Yet the mechanisms that allow the ocean to hide carbon and heat are exquisitely sensitive to density, wind, and stratification—to the very ingredients being altered by climate change. The ocean's gift has never been unconditional. It has been contingent on a circulation pattern that, for millennia, has acted as a reliable conveyor between surface and depth. If that conveyor slows, the ocean still absorbs, still mixes, still breathes, but it does so with a different rhythm, and the atmosphere feels the difference like an unpaid debt coming due.

Chapter 9

Navigating the Future

The ocean's great overturning circulation is not an immutable force, but a fragile system subject to the pressures of human activity. We investigate the subtle statistical tremors that might herald a collapse and explore the stark policy choices required to preserve the climate's global stabilizer before the point of no return.

9.1 Signals in the Noise

A thing can be weakening and yet still feel vigorous. Anyone who has watched an elderly door-spring still swing shut—just a little more lazily each year—knows the trick. The Atlantic overturning circulation, often discussed through the Atlantic Meridional Overturning Circulation (AMOC), lives in that uncomfortable space between endurance and fragility. It is a moving target: winds change, storms rearrange surface waters, winters bite harder or softer, and eddies spin like weather systems in the sea. Against that background, the question that keeps scientists awake is not simply whether the AMOC is changing, but whether it is losing *stability*.

Stability has a specific meaning here. A stable system can be pushed and then return. A less stable system can be pushed and return more slowly; a system close to a threshold might not return at all, instead slipping into a different mode. The core difficulty is that the ocean is noisy even when nothing fundamental is shifting. If you only look at the size of the fluctuations, you can talk yourself into almost any story. The art, and increasingly the mathematics, lies in separating fluctuations that are merely *weather in the ocean* from fluctuations that carry the fingerprints of an approaching tipping point.

The most famous of those fingerprints goes by a deceptively gentle name: *critical slowing down.* The intuition is simple enough for a dinner party, but subtle enough to fuel a decade of methodological disputes. Imagine the AMOC as a system that is constantly being nudged—by gusts of heat from the atmosphere, by bursts of freshwater from rain and river runoff, by meltwater from ice, by the random swirl of eddies. Each nudge is a perturbation. When the AMOC is robust, it resists these nudges and snaps back. When it is losing stability, the same nudges linger. The system takes longer to recover from each shove. The ocean begins to "remember" its recent past more strongly.

That memory is measurable. If a time series has strong memory, today resembles yesterday more than usual. Statisticians call one common measure *lag-1 autocorrelation*: the correlation between a variable and itself one time step later. Autocorrelation can be a sign that the system is slow to return to its preferred state. A second clue is variance. If recovery is sluggish, the system can wander farther from its typical value before it is pulled back, so fluctuations grow larger. Increased autocorrelation and increased variance, taken together, are often treated as a pair of early warning signals.

A conceptual picture makes this almost too easy to believe. Many natural systems can be imagined as a ball resting in a valley. The ball is the state of the system; the valley is the set of forces pushing it toward equilibrium. A deep, steep valley represents strong stability: roll the ball aside and it comes quickly back. A shallow, flattened valley represents weaker stability: the same push sends the ball farther away, and it drifts back slowly, wobbling as it goes. Near a tipping point, the valley can become so flat that a small push can carry the ball over a ridge into a different valley—a new regime.

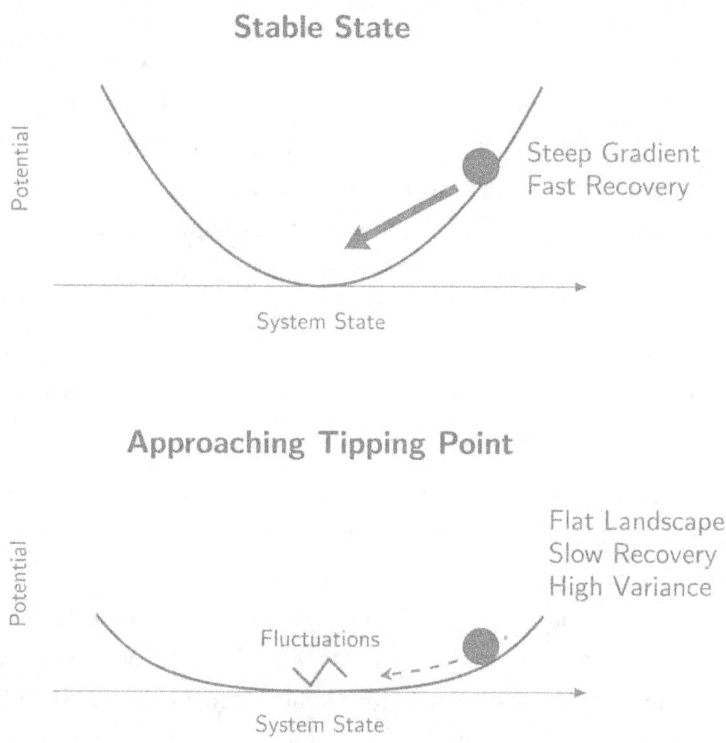

If this were the whole story, the ocean would politely warn us with neat statistical trends: variance up, autocorrelation up, tipping point ahead. Real oceans are not polite. They hide their intentions behind changing measurement practices, shifting winds, seasonal cycles, and the fact that "the AMOC" is not a single dial one can read directly. Scientists infer it from moorings, from sea surface temperature patterns, from salinity, from sea level, from currents at particular chokepoints, and from paleoclimate proxies. Each inference carries its own noise, its own gaps, and its own assumptions.

A historical episode helps clarify why the noise is not an inconvenience but the main plot. In the late 1960s and 1970s, oceanographers noticed something peculiar in the North Atlantic: a vast lens of fresher-than-usual water spread and persisted across subpolar regions, later dubbed the Great Salinity Anomaly. It was not a single storm or a single river flood. It was an ocean-scale rearrangement, likely connected to changes in Arctic freshwater export and sea ice. Freshwater matters because it makes seawater lighter, and in the North Atlantic the formation of dense water is one of the processes feeding the deep branch of the overturning circulation. When the surface is anomalously fresh, winter cooling has a harder time making surface waters dense enough to sink. In other words, the Great Salinity Anomaly was a real-world perturbation to one of the levers that can brake the AMOC.

What makes the episode so instructive is not that it "proved" an impending collapse—it did not—but that it demonstrated how large, persistent, and spatially complicated a perturbation can be without any dramatic, one-time "switch" being thrown. It also illustrated the ocean's long memory. An anomaly introduced in one region can circle the subpolar gyre and return years later, like a message in a bottle with no guarantee of when it will wash ashore. If you were trying to detect early warning signals in a record that spans only a few decades, an event like this can look like a trend, or can mask a trend, depending on where you measure and when you start counting.

The statistics of early warning signals were originally developed in ecology and dynamical systems theory, then carried into climate science with great excitement. Marten Scheffer and colleagues (2009) popularized the idea that critical slowing down could be a generic precursor to critical transitions. The appeal is obvious: if you cannot see the ridge in the landscape directly, perhaps you can infer that the landscape is flattening by watching how the system jiggles. In practice, researchers often compute autocorrelation and variance within a sliding window of time and ask whether those metrics increase. They also look for “flickering”: brief, reversible excursions into a different regime that become more frequent as the boundary between regimes weakens.

In the AMOC context, the excitement surged when studies began to report early warning signals in indices derived from observations. Niklas Boers (2021) examined several independent AMOC-related indices built from sea surface temperature and salinity fields, reporting statistical patterns consistent with decreasing stability. Later, Peter and Susanne Ditlevsen (2023) used a different statistical approach to infer a potential critical transition, drawing intense attention well beyond the scientific community. Those results landed in a world that was primed for a dramatic headline, because the AMOC is easy to imagine as a switch: “on” means a familiar climate for the North Atlantic rim; “off” means a different one.

Yet scientific arguments are not settled by how imaginable a story feels. They are settled by whether the signal survives contact with alternative explanations. A particularly pointed critique came from Xianyao Chen and Ka-Kit Tung (2023), who argued that some early warning signals could be artifacts of changing observational coverage. This sounds technical, but it is deeply human: the ocean was not always watched with the same intensity. As measurement networks expanded and datasets assimilated more observations, the apparent variability in reconstructed fields could increase even if the real ocean variability did not. A time series built from such changing coverage can display rising variance simply because the “camera” got better at seeing fluctuations.

This is the kind of problem that makes climate data analysis feel less like reading a thermometer and more like restoring an old painting.

Are you seeing a new crack in the paint, or did someone finally remove the grime? When variance rises, you must ask whether the ocean truly began to swing more wildly, or whether the observing system began to resolve motions that were always there.

Autocorrelation can also mislead. A slow trend imposed from outside—say, steadily increasing freshwater input—can create the appearance of higher autocorrelation even if the internal stability has not changed. Seasonal cycles can leak into the analysis if not removed carefully. A time series that mixes different dynamics (for example, a surface-temperature-based AMOC index that responds both to ocean circulation and to direct atmospheric forcing) can show memory changes that reflect the atmosphere rather than the deep circulation. Even the choice of time scale matters. The AMOC contains variability from weeks to decades. "Lag-1 autocorrelation" depends on what "one lag" means: one month, one year, one decade. A system can be sluggish on one scale and lively on another.

One way to appreciate the challenge is to imagine trying to predict a person's health from the bounce in their walk while they are also walking on a moving ship. The bounce changes because the knees are weakening, but it also changes because the deck is pitching, because the person is carrying luggage, because they are tired, because the shoes changed. The AMOC's "walk" is observed from a ship that is itself moving: the climate system.

Because of these pitfalls, researchers increasingly insist on a kind of triangulation. If critical slowing down is real, it should appear across multiple independent lines of evidence, not just one index built from one dataset. It should be robust to plausible choices about detrending, filtering, window length, and statistical model. It should be detectable in physics-based climate model experiments designed to approach a tipping point under controlled forcing, and those experiments should help interpret what the statistics mean in a messy world.

It should also, ideally, connect to mechanisms: for the AMOC, mechanisms include the salt-advection feedback (where a weaker AMOC transports less salty water northward, potentially making the North Atlantic fresher and further weakening sinking), and the competing influences of winds, mixing, and Southern Ocean upwelling that can maintain an overturning even when northern sinking is reduced.

There is a quiet but crucial distinction here between *predicting* a tipping point and *monitoring* the approach to one. Early warning signals do not typically give a date. They offer a statement about changing stability: the system may be losing its ability to recover. That is closer to a doctor's warning about elevated risk than to a countdown timer. And like medical risk, it is easy to misunderstand. People want certainty: either the AMOC will collapse by a particular year or it will not. The statistics rarely grant that satisfaction, especially when the data record is short relative to the ocean's slow modes.

The shortness of the direct record is not a minor detail; it is a governing constraint. Continuous, mooring-based measurements designed to estimate the AMOC strength across a section of the Atlantic (such as the RAPID array at about 26.5°N) began in the early twenty-first century. That is a triumph of engineering and international cooperation, but it also means that the most direct AMOC observations are younger than many of the people analyzing them. In that brief time, the measured overturning has shown substantial year-to-year variability, including notable dips and rebounds. Those swings are scientifically valuable because they reveal what the AMOC can do on short time scales. They are also statistically treacherous because a handful of big excursions can dominate a trend estimate or an early-warning metric.

Here the ocean plays another trick: a system can show increased variance for reasons that have nothing to do with tipping. If winds become more variable, surface forcing becomes more variable. If deep convection becomes more intermittent because winters fluctuate between harsh and mild, variance can rise. Those changes may matter for climate impacts, but they are not the same as a loss of stability in the underlying dynamical sense. An early warning indicator that responds to both can raise alarms that are real in one sense (the climate is becoming more erratic) and ambiguous in another (the AMOC's stability landscape may or may not be flattening).

There are statistical techniques designed to cope with this ambiguity. Some approaches test whether variance and autocorrelation rise together in a way consistent with a slowing linear recovery rate. Others fit simple stochastic dynamical models (often variants of the Ornstein–Uhlenbeck process) and estimate a "return rate" parameter that directly encodes how quickly the system relaxes after perturbation. Some methods compare observed trends in indicators to what would be expected under a null hypothesis that includes known features of the observing system. Others use spatial patterns rather than single time series: if the AMOC weakens, it leaves fingerprints in temperature and salinity fields across the Atlantic. Caesar and colleagues (2018) drew attention to such a fingerprint in sea surface temperature patterns, including a relative cooling in the subpolar North Atlantic that stands out against broad warming. Spatial fingerprints are not immune to confounding influences, but they offer a different angle: rather than asking whether a single index jiggles differently, they ask whether the Atlantic looks increasingly like a circulation system in a different mode.

Paleoclimate adds yet another angle, though at the cost of precision. Sediment cores, isotopes, and other proxies suggest that the overturning circulation has shifted abruptly in the past, sometimes associated with freshwater injections and reorganizations of North Atlantic convection. Those records are the reason tipping points are taken seriously: they show that abrupt change is physically possible.

But proxies do not provide the clean, evenly spaced time series that early warning statistics prefer. They are irregular in time, uncertain in calibration, and often represent local conditions that must be interpreted as part of a larger circulation story. The ocean does not leave us a tidy logbook.

The public debate around AMOC early warnings therefore has a curious double character. On the one hand, there is a solid physical basis for concern: the AMOC depends on density differences that can be eroded by warming and freshening, and it has feedbacks that can amplify change. On the other hand, there is a technical, sometimes esoteric fight over whether the statistical fingerprints are genuine or drawn by the pencil of the data-processing pipeline. Even within the scientific literature, high-profile claims can be followed by corrections and reanalyses, as happened with a formal correction to Ditlevsen and Ditlevsen's 2023 paper published in 2025. That is not a sign of weakness in science so much as a sign of stress-testing: when claims have big implications, the scrutiny becomes intense, and small assumptions stop being small.

It helps to remember what a "tipping point" means in practice. It is not necessarily a dramatic Hollywood shutdown where the Gulf Stream stops and Europe freezes overnight. The Gulf Stream is largely wind-driven along the U.S. East Coast; the AMOC is the broader overturning that includes deep return flow and northward heat transport. A large weakening of the AMOC could unfold over decades and still be societally disruptive: shifting rainfall belts, raising sea level along parts of the North American Atlantic coast, changing storm tracks, stressing marine ecosystems, and altering regional extremes. A system can be near a threshold in terms of stability while its observable impacts arrive as a mixture of gradual drift and occasional jolts.

So what does "signals in the noise" look like when done responsibly?

It looks like humility paired with persistence. It looks like multiple indices, carefully compared. It looks like sensitivity tests that ask whether the warning appears only under one convenient preprocessing choice. It looks like methods that explicitly account for changing data coverage and measurement error. It looks like comparisons to model experiments where freshwater forcing is increased in a controlled way, allowing researchers to learn which statistical signatures truly precede a known loss of stability. It looks like a willingness to say, "this is suggestive," rather than, "this is certain."

And it looks like an insistence on improving observation itself, because the cleanest way to reduce statistical ambiguity is to watch the ocean better. More moorings, more floats, better satellite constraints on sea level and surface salinity, and sustained funding for the unglamorous work of calibration and continuity all translate into sharper inference. Early warning signals are not magic; they are a way of extracting meaning from limited information. In a system as vast as the Atlantic, the difference between a false alarm and a real warning can hinge on details as mundane as whether a dataset gained new ship tracks in the 1970s or whether an instrument drifted slightly over a decade.

The deeper reason early warning signals matter is psychological as much as scientific. Humans are not good at believing in slow loss of stability. We respond to crises, not to flattening valleys. The statistics of critical slowing down offer a language for that flattening: a way to say that resilience is being spent. Even if the ocean never hands us a single, decisive number that reads "collapse imminent," the act of listening for these signals changes how we understand responsibility. It shifts attention from spectacular endpoints to the quieter erosion of the system's capacity to recover—the moment when a perturbation stops being a temporary wobble and starts becoming a new normal.

9.2 The Mitigation Imperative

A weakening circulation can feel like a far-off technical problem until it is translated into a ledger the ocean keeps with ruthless honesty: heat and freshwater in, density out. The thermohaline circulation is powered by differences in seawater density, and density is determined mainly by temperature and salinity. Warm water is lighter than cold; fresh water is lighter than salty. Greenhouse gas emissions tilt both levers in the same destabilizing direction. They warm the surface ocean directly, making it harder for winter cooling to produce dense water. They also accelerate the melting of land ice and increase high-latitude precipitation and river runoff, freshening the surface and further resisting sinking. That is the physical bridge between an emissions pathway drawn on a policy graph and the probability of a circulation failure drawn in the ocean.

The bridge is not a single plank. Emissions set the rate of atmospheric warming; warming sets the pace of ice loss and rainfall change; those set the freshwater and heat delivered to the North Atlantic; those set the density structure; and that structure determines whether deep water formation regions can still do their job. None of these links is perfectly linear, which is exactly why mitigation matters. Linear problems allow you to bargain: a bit more warming yields a bit more impact. Nonlinear systems have cliffs. Mitigation is the effort to keep the system away from the part of the landscape where small additional forcing buys you a large loss of stability.

A useful way to think about the AMOC is as a conveyor with a sensitivity knob. Turning the knob does not stop the machine immediately; it changes how vigorously it runs and how easily it can be perturbed. Greenhouse gases turn the knob in a consistent direction. The ocean responds with both speed and delay, like a heavy flywheel. Some changes arrive quickly—surface warming, altered storm tracks, shifts in evaporation and rainfall. Others arrive late, because the ocean stores heat and because ice sheets do not melt on human time scales in a smooth, obedient way. That mix of immediacy and inertia can make mitigation feel psychologically unsatisfying: the benefits do not always arrive on the same schedule as the effort.

The Intergovernmental Panel on Climate Change (IPCC), in its Sixth Assessment Report Working Group I (2021), judged that an AMOC weakening over the twenty-first century is very likely under all the standard future scenarios. There is a further twist that can surprise non-specialists: the *rate* of weakening in many models is assessed as approximately independent of emissions scenario, at least through much of the century, even though the *amount* of warming differs substantially. This is not a loophole that makes mitigation pointless; it is a warning about lags and mechanisms. If the circulation is already being pushed by a combination of warming and freshwater input, it may keep declining for a time even in a world that begins to stabilize temperature, much as a slowing bicycle still coasts forward after you stop pedalling. In some model experiments where warming stabilizes or emissions cease, the AMOC can continue to decline for roughly five to ten years after global mean temperature stabilizes and then recover over centuries. That is an uncomfortable timeline for politicians and a familiar one for geophysicists: the ocean does not hurry, but it also does not forget.

If mitigation does not immediately reverse a weakening, why call it an imperative rather than an aspiration? Because the crucial question is not whether the AMOC weakens a little; it is whether the system is pushed across a threshold where feedbacks amplify the change and recovery becomes slow, uncertain, or impossible without a large reversal of forcing. Those thresholds are commonly framed in terms of a freshwater forcing that overwhelms the salinity-driven density contrast needed for sinking in the North Atlantic. In the simplest terms, mitigation is a way of limiting the freshwater and heat perturbation that the ocean must absorb. It lowers the chance that natural variability plus anthropogenic forcing add up to the shove that carries the system into a new regime.

The language of climate policy is full of acronyms, but behind them lie choices about physics. The Shared Socioeconomic Pathways (SSPs) used in recent climate projections are storylines about population, energy, land use, and technology that translate into emissions trajectories. At one end, a low-emissions pathway such as SSP1-2.6 produces substantially less warming than a high-emissions pathway such as SSP5-8.5. For the AMOC, the difference between these pathways is not merely a different endpoint temperature. It is a different trajectory of high-latitude freshening, a different stress on Greenland's margins, and a different risk of crossing a stability boundary. Even if many models show a weakening across scenarios, the high-emissions world is more likely to push the system into the zone where the statistical early warning signs become more than an academic debate.

A concrete story makes the chain from emissions to ocean instability less abstract. In the 1990s and early 2000s, Greenland's outlet glaciers became household words among glaciologists: Jakobshavn Isbræ on the west coast, Helheim and Kangerdlugssuaq on the east. Satellite observations revealed acceleration and retreat that were faster than many people expected from an ice sheet that had been treated, for decades, as slow and conservative. The acceleration was not only a matter of air temperature. Warm Atlantic-origin waters intruding into fjords can melt glacier fronts from below, thinning them and reducing the buttressing that slows their seaward flow. This is a vivid example of coupling: the ocean can speed up ice loss, and ice loss can then freshen the ocean. It is not hard to see how a feedback loop could arise, even if the loop's strength and timing are still actively researched.

The historical anecdote here is not a single dramatic day but a shift in scientific posture. During the twentieth century, the dominant picture of Greenland mass balance was shaped by sparse measurements and the sense that ice sheets move at geological pace. The satellite era changed that. The public learned a new kind of climate narrative: not merely the gradual rise of a global average temperature, but sudden reorganizations in complex systems. That narrative has a double edge. It can be used responsibly, as a motivation for risk management under uncertainty. It can also be used recklessly, by turning every acceleration into a countdown. The honest middle ground is that warming increases the probability of rapid ice dynamics and increases the magnitude of meltwater input over time; both of those are relevant to the AMOC even if they do not translate into a simple collapse date.

Mitigation also matters because of the way the ocean partitions heat between surface and depth. A warmer world does not only warm the air; it loads the ocean with heat, and the ocean stores most of the excess energy trapped by greenhouse gases. That heat storage has two consequences for circulation risk. First, it increases stratification: a warmer surface layer sits more stably atop colder deep water, making vertical mixing and sinking harder. Second, it means that the forcing persists even if emissions decline. Carbon dioxide remains in the atmosphere for a long time, and the heat already in the ocean does not vanish when a policy is signed. A circulation that weakens in response to accumulated heat and freshwater may therefore continue weakening briefly even as mitigation begins to bite, before any recovery becomes possible.

The policy world has tried to translate this physics into timelines. The IPCC Working Group III Summary for Policymakers (2022) assessed that pathways limiting warming to 1.5°C with no or limited overshoot (with greater than 50% probability) and pathways limiting to 2°C (with greater than 67% probability) involve global greenhouse gas emissions peaking between 2020 and at the latest before 2025, followed by rapid and deep reductions. Those dates were not chosen for rhetorical punch; they arise from carbon budgets and the inertia of the energy system. The ocean does not care whether the peak was politically difficult. It cares about how much additional heat and freshwater forcing it must integrate over decades.

The Paris Agreement's long-term temperature goal—holding warming well below 2°C and pursuing efforts to limit it to 1.5°C—can feel remote from an oceanographer's daily work. Yet for the AMOC it is directly relevant, because the difference between those goals is not only an average temperature difference. It is a difference in the likelihood of stronger Greenland melt, the intensity of the hydrological cycle, and the background stratification against which winter storms attempt to ventilate the subpolar North Atlantic. When policymakers argue about fractions of a degree, the ocean hears it as a debate about density.

Risk also has a geographic signature. Freshwater forcing is not uniform. What matters most for North Atlantic deep water formation is the local balance: precipitation minus evaporation, river input, sea ice melt and export, and the meltwater flowing off Greenland into boundary currents and subpolar basins. A global emissions pathway affects all of these, but not equally. That is one reason why the same global warming level can produce different outcomes depending on patterns of warming and freshwater delivery. Climate models differ in these patterns, and that difference contributes to uncertainty about thresholds. Mitigation reduces risk partly by reducing the global mean change, and partly by limiting the extremes and regional anomalies that can push critical regions over the edge.

A detail often missed in popular discussions is that there is more than one way to "brake" the overturning. Freshening is an obvious one. Warming is another. But winds and mixing can either oppose or reinforce the braking. Stronger Southern Ocean winds, for example, can sustain upwelling that keeps an overturning circulation operating even when northern sinking is reduced, while changes in vertical mixing can alter how readily dense waters are formed and exported. This matters for mitigation because it reminds us that uncertainty does not simply mean "we do not know." It means multiple mechanisms compete, and the balance between them can shift with emissions-driven climate change. Mitigation is a way of keeping the forcing small enough that these competing mechanisms are less likely to be overwhelmed by one dominant destabilizing influence.

The temptation is to translate mitigation into a single promise: stabilize temperature and the AMOC will stabilize. The ocean is more complicated. In some model experiments, as noted by IPCC Working Group I (2021), the AMOC continues declining for years after global temperature stabilizes. The deep ocean adjusts slowly, and density structure can keep evolving even without further surface warming. The promise of mitigation is therefore not an instant cure but a reduction in cumulative risk: a smaller chance of crossing a threshold, a smaller magnitude of weakening, and a larger probability that recovery remains physically available on long time scales.

There is also a moral asymmetry in threshold problems. A strong circulation can be weakened quickly by a pulse of freshwater, but rebuilding the dense-water formation machinery can take far longer, because it requires removing freshwater, cooling the surface, and re-establishing the salinity and temperature patterns that support sinking. In the ball-in-a-valley picture, mitigation is not simply pushing the ball uphill; it is preventing the valley itself from flattening. Once flattened, even a strong shove in the opposite direction may not restore the old landscape quickly.

A concrete analogy with a human timescale helps. Consider a coastal city that has gradually filled in marshland over a century to build houses. The city might endure storms for decades with only modest flooding. But each additional development removes a little more sponge-like capacity. A storm that would once have been absorbed becomes a disaster. The difference between a careful plan that preserves wetlands and a careless one that does not may not show up for years. When it does, it shows up as a threshold: the day the water reaches the electrical substations, the day the hospital loses power. The AMOC is not a city, but it shares this feature: resilience can be spent quietly, and then the system reacts sharply to a perturbation that would previously have been routine.

The question then becomes how to connect emissions trajectories to circulation probabilities in a way that is honest about uncertainty. Climate science has learned to avoid the trap of false precision. Rather than stating that an AMOC collapse will occur at a specific warming level, assessments more often say that weakening is very likely, that abrupt collapse before 2100 is not expected with medium confidence (as in IPCC Working Group I), and that uncertainty is high because of limited observations and potential missing processes. That posture can sound noncommittal, but it carries a clear implication for mitigation: if a risk has catastrophic potential and a fat-tailed uncertainty—meaning that low-probability outcomes may still be severe—then reducing the forcing that pushes toward that tail is rational even when the median outcome is less dramatic.

Recent research has sharpened one piece of the story in a way that matters for mitigation: the role of freshwater forcing in reconciling models with observations. Analyses reported in the mid-2020s, including a Nature Geoscience study in 2024, argued that accounting for stronger subarctic freshening and meltwater input can produce a modeled AMOC slowdown more consistent with observation-based indications of weakening since the mid-twentieth century. This is not a small technical adjustment. If models have underestimated historical freshwater forcing, they may also underestimate how sensitive the AMOC is to future meltwater. Mitigation becomes more urgent in that framing because it is the primary lever for limiting future ice loss and hydrological intensification.

The UN Environment Programme's Emissions Gap Report (2025) added another policy-relevant layer: it emphasized that current policies and national commitments still imply warming well above Paris goals and warned that a 1.5°C overshoot is very likely within the next decade without rapid additional reductions. Whether one focuses on 1.5°C or 2°C, the point for circulation risk is that delaying mitigation is not neutral. It commits the ocean to additional heat uptake and increases the probability that freshwater forcing crosses levels that are difficult to reverse.

There is a final, quietly unsettling implication of the AMOC's inertia: mitigation is most valuable when it prevents a threshold from being crossed, not when it tries to repair the system after the fact. Once a circulation has weakened substantially, societies may be tempted to rely on later technological fixes—large-scale carbon dioxide removal, or even more speculative climate interventions—to claw back stability. But the physical system does not guarantee that reversing emissions yields a smooth reversal of the circulation, and the timescales involved may stretch beyond political lifetimes. Mitigation is not merely about avoiding some distant, dramatic collapse; it is about staying in a regime where the climate system's slow components still behave as slow components rather than as systems primed for abrupt reorganization.

The ocean's message, when translated into the language of emissions, is starkly practical. Every ton of greenhouse gas added to the atmosphere adds heat to the ocean and strengthens the forces that freshen and stratify the North Atlantic. The AMOC may not respond with a spectacular lurch on command, but it responds with accounting: weakened sinking here, altered heat transport there, a little less ability to recover from the next perturbation. Mitigation is the choice to stop increasing the load on that accounting system while there is still room for resilience to matter.

9.3 Decisions at the Precipice

A society can be rational and still be unprepared. The trouble begins when the biggest dangers are not the most likely ones. Most people buy home insurance not because they expect a fire, but because the cost of being wrong is ruinous. Yet even this familiar logic falters when the hazard is both uncertain and collective. With the Atlantic overturning circulation, the uncomfortable possibility is not that the AMOC *must* collapse soon, but that it *could* weaken far enough, fast enough, to reshape regional climates and coastlines in ways that are hard to reverse. It is a problem of governance under deep uncertainty: acting decisively when the probability is contested, the consequences are unevenly distributed, and the costs of prevention are paid now while the benefits accrue later.

The temptation is to ask for a single number. What is the probability of an abrupt AMOC collapse by 2100? The IPCC's Sixth Assessment Report Working Group I (2021) judged, with medium confidence, that an abrupt collapse before 2100 is not expected. That phrasing is careful. "Not expected" is not the same as "impossible." Medium confidence is not high confidence. The assessment also reflects the limits of current models and the shortness of direct observations. This is a situation in which science provides boundaries rather than certainties: weakening is very likely; abrupt collapse this century is assessed as unlikely; uncertainties remain substantial.

When people hear "unlikely," they often translate it into "ignore it." That is a reasonable everyday heuristic when the stakes are modest. It is a dangerous heuristic when the stakes are systemic. In financial crises, in pandemics, in nuclear safety, a low-probability event can be the event that defines an era. The ethical challenge in climate tipping risks is that the people who pay to reduce the risk are not necessarily the same people who would suffer the most if the risk is realized, and the people who would suffer may not yet be born. Ordinary cost–benefit thinking, which discounts future harms and looks for a best estimate, strains under that weight.

The AMOC makes this strain visible because it sits at an awkward intersection of plausibility and consequence. A drastically weakened overturning would likely shift rainfall patterns, affect storm tracks, alter marine ecosystems, and raise sea level along parts of the North American Atlantic coast through dynamic changes in ocean circulation. Even without a dramatic "shutdown," a large weakening could still be socially destabilizing. Yet the evidence for a near-term tipping point is contested. Studies such as Boers (2021) have reported early warning signals consistent with a loss of stability in observation-based AMOC indices, while critiques such as Chen and Tung (2023) have argued that such signals could arise from changes in observational coverage and other artifacts. Ditlevsen and Ditlevsen (2023) drew intense attention by inferring a possible critical transition window from statistical analysis of proxy data, and the fact that a formal correction was later published in 2025 is a reminder that high-impact claims can hinge on technical details. A policymaker looking at this literature does not see a single marching order; they see an active argument.

There is a simple moral danger in that argument: to mistake uncertainty for safety. A system can be poorly measured and still be at risk.

Uncertainty cuts both ways. It can mean the danger is overstated; it can also mean it is understated. Climate models may omit processes that matter for freshwater forcing, ice—ocean coupling, or small-scale mixing. Observations may be too short to capture slow modes. Paleoclimate records may not resolve the mechanism with the clarity one would like. In such circumstances, a precautionary approach is not a slogan but a structured response: choose actions that are robust to a range of plausible futures rather than optimal for a single best guess.

A historical analogy clarifies what "robust" can mean without reducing it to abstract philosophy. In the late 1980s, scientists discovered a seasonal "ozone hole" over Antarctica. The basic chemistry of chlorofluorocarbons (CFCs) was understood, but the precise evolution of ozone depletion was uncertain, and the economic uses of CFCs were widespread. The Montreal Protocol (1987) is often celebrated because it worked, but it is also a case study in decision-making under uncertainty. Governments did not wait for perfect prediction of the exact size and timing of ozone recovery. They acted on a clear mechanism, credible evidence of harm, and the availability of substitutes. Years later, the slow improvement of stratospheric ozone vindicated the decision, but the decision itself had to be made in a fog of incomplete knowledge.

The AMOC is harder. There is no single chemical to ban and no single substitute to deploy. Mitigation involves reshaping energy systems, land use, and industrial processes. Adaptation involves infrastructure and planning. Even the harms are more diffuse: not a single measurable UV index spike but a rearrangement of climate statistics. And still, the structure of the dilemma is similar. The question is not whether one can forecast the exact moment when the system crosses a threshold. The question is whether one has enough mechanistic understanding to justify treating the risk as a meaningful part of climate policy.

Mechanism offers some firmness. The AMOC depends on density contrasts that can be weakened by warming and freshening. Feedbacks such as salt-advection can amplify change. Freshwater forcing from Greenland melt and high-latitude precipitation can interfere with deep water formation. Recent mechanistic work discussed in Nature in 2024 has suggested that complete collapse may be harder than some simplified pictures imply, because Southern Ocean winds and associated upwelling can sustain an overturning even under strong forcing. That is reassuring in one narrow sense, but it also contains an unsettling message: a system can avoid "complete" collapse and still become weak enough to cause major impacts. The very definition of collapse becomes contested. Some studies use an operational threshold in Sverdrups (Sv) of overturning strength; others emphasize qualitative regime change. In policy terms, this matters because waiting for a dramatic, unambiguous collapse signal is like waiting for a bridge to fall rather than monitoring for cracks.

The decision problem is further complicated by communication. Scientists have learned, sometimes painfully, that dramatic tipping-point language can backfire. A high-profile paper can be summarized in headlines as a countdown to catastrophe. A critique or correction can then be summarized as "scientists were wrong," even if the underlying risk remains. This is not merely a public-relations issue. It shapes trust, and trust shapes whether societies can coordinate long-term action. The tipping-points literature itself contains cautionary tales, including corrections to prominent commentary pieces, which serve as reminders that urgency must be paired with careful claims.

So how should a society weigh a low-probability, high-impact ocean event? A practical answer begins by admitting that the decision is not only about the AMOC. Climate policy is already justified by high-probability harms: heat extremes, sea level rise, ecosystem loss, crop stress. The AMOC functions as a risk amplifier in the portfolio: a possibility that adds fat-tail outcomes—rare but severe scenarios that standard planning tends to miss. In finance, fat tails are the reason you do not manage risk by averages. In climate, they are the reason "expected value" calculations can mislead.

Economists typically compare the costs of mitigation with the discounted value of avoided damages. Discounting is the practice of valuing future costs and benefits less than present ones, often because money invested today grows over time or because people prefer benefits now. Discounting makes sense for many decisions, but it becomes ethically loaded when future harms involve lives, cultures, and ecosystems that cannot be repaid with interest. A high discount rate can make it appear rational to accept large future risks in exchange for small present savings. A low discount rate can make aggressive mitigation appear overwhelmingly justified. The AMOC intensifies this dispute because the most dramatic impacts, if they occur, may unfold over decades and centuries, while the emissions that influence the risk are being decided now.

One way out of this ethical tangle is to treat certain outcomes as unacceptable regardless of their precise probability, much as societies treat catastrophic engineering failures. Nuclear plants, aircraft, and dams are not designed by maximizing average performance; they are designed with safety margins, redundancy, and conservative assumptions. That approach does not eliminate accidents, but it changes the decision frame. The relevant question becomes: what safety margin is appropriate given the stakes and the uncertainty?

Applied to the AMOC, a safety-margin approach would emphasize actions that reduce forcing on the system (emissions cuts), improve monitoring (so that warnings are clearer), and build resilience to plausible impacts (coastal planning, agriculture and water management, ecosystem protection). It would also resist the false comfort of binary thinking. The world does not have to choose between "no problem" and "instant collapse." Many of the harms associated with the AMOC are graded: incremental weakening changes probabilities of extremes, shifts seasonal patterns, and alters regional sea level. Robust policy prepares for these graded impacts while also aiming to keep the system away from threshold-like regions of its stability landscape.

A second way out is to acknowledge that the decision is unavoidably political, and then insist on fairness. The emissions that contribute most to forcing are historically concentrated in certain countries and sectors, while many of the communities most exposed to climate disruption have contributed least. If an AMOC weakening raises sea level along parts of the U.S. East Coast, the costs fall on homeowners, renters, insurers, and local governments, with disproportionate impacts on lower-income communities who have fewer options to relocate or rebuild. If rainfall patterns shift in the Sahel or Amazon, the consequences can be existential for farmers and for ecosystems that regulate global carbon. Ethical decision-making therefore requires not only reducing the risk but distributing the burden of risk reduction and adaptation in a way that does not deepen existing inequalities.

The AMOC also poses a question about the role of "last resort" technologies. When mitigation is delayed, interest grows in interventions that might reduce global temperature quickly, such as solar geoengineering (for example, injecting reflective particles into the stratosphere). The ethical attraction is easy to see: a fast lever for a slow-moving emergency. The ethical hazards are equally stark: uneven regional effects, governance disputes, moral hazard (the temptation to substitute a risky patch for emissions cuts), and the possibility that abruptly stopping such interventions would cause rapid warming. For an ocean circulation risk, there is an additional complication:

even if solar geoengineering cooled the surface, it might not reverse ocean heat content on the timescales relevant to circulation recovery, and it would not directly remove the freshwater already added from ice melt. The intuition that "cooling fixes it" can be dangerously incomplete.

Carbon dioxide removal (CDR) is less physically abrupt but still ethically and economically complex. Large-scale removal could, in principle, reduce atmospheric CO_2 and allow the climate system to cool over time. Yet it requires land, energy, money, and governance. It also carries intergenerational implications: overshoot now, clean up later. In an AMOC context, overshoot is particularly troubling because thresholds can be crossed during the overshoot period even if temperatures later decline. Limiting the magnitude and duration of overshoot therefore matters, not only for coral reefs and glaciers but also for circulation stability.

The practical governance question becomes: how should early warning signals influence action? An early warning is not a prophecy; it is an indicator of changing stability. The right analogy is not an alarm clock but a medical scan. A scan that shows a potential problem does not guarantee a diagnosis, but it can justify preventive measures, closer monitoring, and lifestyle changes that have broader benefits anyway. In climate policy, emissions reductions have broad benefits beyond the AMOC: improved air quality, reduced heat stress, avoided ecosystem loss. Increased ocean observation has broad benefits beyond tipping risk: better seasonal forecasts, improved storm prediction, safer shipping, better fisheries management. Actions that are beneficial across many futures are the backbone of robust decision-making.

A final, human dimension is easy to miss in technical debates about Sverdrups and autocorrelations. Decisions about low-probability catastrophes force societies to confront what kind of solidarity they are willing to practice across distance and time. The ocean's circulation is a global commons in a literal physical sense: it moves heat and salt between hemispheres, linking weather in one region to ice in another. The economic system is also a commons in a social sense: supply chains, migration pressures, food prices, and insurance markets do not respect borders. The ethics of AMOC risk is therefore not only a question of saving a current pattern of climate for its own sake. It is a question of whether societies will treat the stability of the Earth system as a shared inheritance, worth protecting even when the threat arrives as a probability distribution rather than as a siren.

The most defensible stance is neither fatalism nor melodrama. It is to treat the AMOC as a component of climate risk that deserves the same sober engineering mindset applied to other high-stakes systems: invest in measurement, reduce the forcing you control, prepare for impacts you cannot fully prevent, and avoid gambling on the hope that uncertainty means safety. In a world already committed to some change, the difference between a manageable future and a brittle one can hinge on whether resilience is preserved while it still exists to be preserved.

www.ingramcontent.com/pod-product-compliance
Lightning Source LLC
LaVergne TN
LVHW022207200926
848481LV00006B/1872
* 9 7 9 8 8 9 6 6 5 2 7 8 6 *